ICE AGE

*John Gribbin and
Mary Gribbin*

ICE AGE

ALLEN LANE

THE PENGUIN PRESS

ALLEN LANE
THE PENGUIN PRESS

Published by the Penguin Group
Penguin Books Ltd, 80 Strand, London WC2R 0RL, England
Penguin Putnam Inc., 375 Hudson Street, New York, New York 10014, USA
Penguin Books Australia Ltd, Ringwood, Victoria, Australia
Penguin Books Canada Ltd, 10 Alcorn Avenue, Toronto, Ontario, Canada M4V 3B2
Penguin Books India (P) Ltd, 11 Community Centre, Panchsheel Park, New Delhi – 110 017, India
Penguin Books (NZ) Ltd, Cnr Rosedale and Airborne Roads, Albany, Auckland, New Zealand
Penguin Books (South Africa) (Pty) Ltd, 24 Sturdee Avenue, Rosebank 2196, South Africa

Penguin Books Ltd, Registered Offices: 80 Strand, London WC2R 0RL, England

On the World Wide Web at: www.penguin.com

First published 2001

3

Copyright © John and Mary Gribbin, 2001

The moral right of the author has been asserted

Set in 11/14.75 pt Adobe Linotype Minion
Typeset by Rowland Phototypesetting Ltd, Bury St Edmunds, Suffolk
Printed and bound in Great Britain by Clays Ltd, St Ives plc

A CIP catalogue record for this book is available from the British Library

ISBN 0–713–99612–9

Contents

Prologue: *The Ice Age Now*   1

One: *The Victorians' Ice Age*   7

Two: *The Serbian's Ice Age*   41

Three: *Deep Proof*   62

Epilogue: *Ice Ages and Us*   91

Sources: 102

Prologue

The Ice Age Now

By the standards of the geological past, we live in an Ice Age. The world has rarely been as cold as it is today. We don't call it an Ice Age, because not too long ago the world was even colder than it is today – that is what we think of as 'the' Ice Age. Before too long, unless human activities prevent it, the world will cool again, back into the Ice Age proper. Our perspective (the entire history of human civilization) embraces only a short-lived, temporary retreat of the ice, an Interglacial. The succession of relatively long-lived Ice Ages and relatively short-lived Interglacials is now known as an Ice Epoch, and lasts for several million years. It is the story of the discovery of the rhythms of the Ice Epoch, and the implications for life on Earth, that we tell here; but even an Ice Epoch is just a passing phase in the lifetime of a planet that has already been around for more than four billion years.

We think that it is normal to have ice at both poles of our planet. After all, there has been ice there for longer than there has been human civilization. But in the long history of the Earth, polar ice caps are rare, and having two polar ice caps at the same time may be unique. Indeed, it may be the presence of those polar ice caps which has made us human. And although we associate weather with the movement of masses of air around the globe, with high pressure systems bringing settled, dry conditions and low pressure systems bringing wind and rain or snow, as far as climate is concerned great ocean currents are much more important. Those currents carry warm water from the equatorial region to the poles, and the polar regions can only freeze at all if that flow of warm water is obstructed. Today, the South Pole is frozen because a great land mass, Antarctica, lies right over the pole, preventing any ocean currents from reaching it. The conditions around the North Pole are almost a mirror image of this, with a nearly landlocked Arctic Ocean almost surrounded by land masses which make it difficult for water to flow northward to the pole. But the warming power of that flow of water away from the equator is nowhere better seen than in northwestern Europe, where the current known as the Gulf Stream, which 'ought' to be warming the pole, has been deflected eastward by the bulge of Canada and by Greenland to make Britain and its neighbourhood some six degrees, Celsius, warmer than it would otherwise be. The simplest way to envisage what the climate of northwestern Europe 'ought' to be like is to look at any globe of the world, and cast your eye due west from Ireland to Canada. Anybody who has experienced a Canadian winter – at the same latitudes as Ireland – would be in little doubt that we are living in an Ice Age.

Six degrees is a lot of warming. If Greenland were not in the way, and the Gulf Stream could warm the Arctic by six degrees, the thin layer of ice floating on the surface of the Arctic Ocean would disappear, changing the climate of the temperate region of the entire Northern Hemisphere. In fact, just a fraction of the heat carried by the Gulf Stream would be enough to do the job. The changes resulting from Arctic warming would be bigger than you might expect at first sight, and would in many ways be unpredictable, because of the effect of positive feedback. Today, the shiny white surface of the ice covering the Arctic Ocean reflects away incoming solar energy, and helps to keep the polar region cool. Once the ice starts to melt, however, it exposes dark water, which absorbs the incoming solar energy and warms the region still further. If the world cooled for any reason, the feedback would operate in reverse, with dark ocean being covered by shiny ice that reflects away incoming solar energy and helps to keep things cold. If the Arctic icecap were removed, by magic, tomorrow, it would not reform. The world would be quite happy to maintain an ice-free Arctic Ocean. Either state is stable – with or without ice. But you can't have half the north polar icecap; the feedbacks make it an all or nothing choice.

Curiously, this kind of process raises the possibility that one of the first effects of the global warming that is going on at present (which most climatologists explain as being at least partly caused by human activities) could be to cool northwest Europe. The argument runs like this. The warm water flowing on the surface of the ocean up the western side of the North Atlantic (carrying thirty million cubic metres of water every second) is part of a global system of ocean currents which flows all the way from the tropical Pacific, around South Africa's Cape of Good Hope, picking up warmth from the Sun for most of its long journey. This warm water is less dense than the cold water of the deeper ocean, which is why it forms a surface current; but it is increasingly salty, because evaporation carries water away into the air. In the far North Atlantic, where the current is giving up its heat to the winds which blow from west to east at those latitudes, carrying the warmth towards Europe, the current becomes colder and more dense. With the added burden of its high salt content, this makes it sink into the depths, where it returns all the way back to its starting point before welling up again in the North Pacific. The whole system forms a kind of conveyor belt, driven by upside-down convection, pushed by the descending dense, salty water of the North Atlantic. The flow of this 'river' in the ocean is twenty times greater than the flow of all the rivers on all the continents of the Earth put together.

If the Arctic icecap began to melt, though, fresh water would mingle with the surface flow of the Gulf Stream, diluting its saltiness and making it less dense. If this stopped the water from sinking, the push that drives the conveyor belt would be turned off, and the whole flow would stop, with warm water no longer being carried northward from

Panama past Florida and up the North Atlantic. The whole flow could even reverse, with Europe cooling to the temperatures typical of Vancouver, and Vancouver enjoying the kind of climate now found in Ireland. If this claim seems over the top, consider that the rate at which heat is being transported northward by the Gulf Stream today is more than a million billion watts, as if the current were a real conveyor belt covered with a million million one-bar electric fires, all pouring their heat out into the atmosphere. And that is the only reason why it isn't obvious to everyone in Britain, and the rest of northwest Europe, that we are living in an Ice Age now.

But there's more to this story than the change of perspective it provides. Because of the positive feedback involving the shiny ice of the Arctic Ocean, small changes in the way the world is heated can produce effects on climate which seem, to someone not used to such feedbacks, out of all proportion to their size. Computer simulations of the so-called greenhouse effect show that under some circumstances an increase in the amount of carbon dioxide in the air which produces a warming equivalent to an increase of merely 0.0002 per cent in the Sun's energy arriving at the Earth could flip the Arctic Ocean into an ice-free state – and, conversely, a comparably small cooling could flip it back again, if the conditions were just right. Because the real world is much more complicated than the computer models, nobody knows how close the real world is to being changed by such a climatic flip, but the message to absorb is that, thanks to positive feedback, small changes in the heat balance of the Earth can have big effects on climate.

So we live in a relatively warm part of an Ice Age (or Ice Epoch), and the kind of changes needed to plunge us back into full Ice Age conditions can be produced by quite small changes in the heat budget of the Earth. Those items of twentieth century information would have been immensely valuable to a handful of scientists in the Victorian era, struggling to make sense of strange scratches in the rocks of Europe, and of heaps of boulders deposited far from the rock strata to which they belonged.

One

The Victorians' Ice Age

When scientists in northern Europe began to observe the world around them, one of the first things they noticed was the existence of what became known as erratic boulders (or simply erratics) – lumps of rock lying around in places far from the strata to which they belong. The term 'boulder' scarcely does justice to some of these lumps of rock, which can be as big as a house. In many parts of Europe, there are jumbled heaps of rocks and sediment, which have clearly been transported there and dropped by some agency. To seventeenth and eighteenth century geologists, the obvious explanation was that this agency was the Biblical Flood. This idea became incorporated in the concept of catastrophism, which was the received wisdom at the end of the eighteenth century. By then, studies of fossils and geological strata had shown that there had been great changes in the Earth itself and in the forms of life on Earth, all of which had to be explained within a timescale thought by many to go back only a few thousand years. The changes were interpreted as resulting from a succession of catastrophes, the hand of

God at work, with the Biblical Flood just the latest in a long line of such events.

In Switzerland, people saw things differently. Surrounded by mountains, and with evidence before their eyes of the power of glaciers, people automatically assumed that erratics seen far down in the valleys today had been dumped there by glaciers which had since retreated up the mountains. This idea was formalized by a Swiss minister of the cloth, Bernard Kuhn, in 1787, but without raising much interest. The Scot James Hutton, one of the founders of scientific geology, visited the Jura mountains of France and Switzerland, and independently arrived at the same conclusion, which he published in the mid-1790s. Once again, nobody took much notice. But Hutton is a key figure in the story, because he was the first person to promote the idea of uniformitarianism, which says that the Earth has been shaped not by immense catastrophes operating in a short interval of time but by the same processes of wind and weather (and volcanism, and so on) that we see around us today, operating on an immensely long timescale. To a human civilization, of course, an Ice Age would be a catastrophe. But on a planet now known to be more than four billion years old, the occasional Ice Age is part of the routine. It's all a matter of perspective.

But although Hutton's geological work led to a fierce debate between the catastrophists and the uniformitarians in the early nineteenth century, nobody took much notice of the Ice Age idea, which was scarcely more than a footnote to Hutton's main work, until it was taken up and vigorously promoted by another Swiss, Louis Agassiz, in 1837. Even Charles Lyell, the British geologist who took up Hutton's ideas and developed them in the nineteenth century (and who was a major influence on Charles Darwin, who applied a similarly uniformitarian view to evolution) accepted that erratics had been carried to their present locations by water, although he did come up with a variation on the theme which envisaged boulders and sediment being frozen into great icebergs, icy rafts which carried the material to its present location before they melted and the water retreated.

Agassiz was just the man to take up and promote the Ice Age idea – but he came to it by a circuitous route. Born in Motier on 28 May 1807, he was the son of a Swiss pastor and the daughter of a physician, their fifth child, but the first to survive infancy; understandably, his parents were very keen for him to qualify in one of the professions and have a safe, respectable career. Although what he was really interested in was natural history, in order to please his parents (and to ensure their financial support) Agassiz studied medicine in Zurich, before moving on first to Heidelberg and then to Munich, ostensibly still as a medical student but in fact mostly studying natural history. He was an outstanding student, who wrote a major treatise on Brazilian fishes, published in May 1829, the year he received his Doctorate (in natural history) from Munich; a year later, he kept his promises to his parents by obtaining his medical degree, although he never practised. Instead, he worked for a time with the great naturalist Georges Cuvier in Paris, where he started to study the vast collection of fossil fishes in Cuvier's care in the Natural History Museum. Cuvier was a leading catastrophist, and like Cuvier Agassiz came to believe that different forms of life seen in different geological strata were not descended from one another, but that after each catastrophe God had re-populated the Earth with new forms of life. When Cuvier died, of cholera, in 1832 (two weeks before Agassiz' twenty-fifth birthday) the young naturalist saw himself as a disciple with the duty to carry forward the work of the great man. Without financial support, he was unable to stay in Paris, but returned to Switzerland in September 1832, where he became Professor of Natural History at a new college being started in the town of Neuchâtel, with an associated museum where he could keep his collections. It was there that he completed his epic study of fossil fishes, becoming a world-renowned authority on the subject, and President of the Swiss Society of Natural Sciences.

Meanwhile, interest in the Ice Age idea had been stirring in the mountains of Switzerland. A convenient place to pick up the story is with Jean-Pierre Perraudin, a mountain-eer who lived in the Val de Bagnes. Perraudin wasn't a mountaineer in the modern sporting sense, but a man who made his living in the mountains (specifically, by hunting chamois), and became convinced, from the scarring of the rocks, as well as the evidence of the erratics, that the glacier which now occupied only the higher end of the valley where he plied his trade had once filled the entire valley. In 1815, Perraudin drew this idea to the attention of the geologist Jean de Charpentier, who was in charge of the salt mines at Bex, in the Rhône valley. Charpentier had started life, in 1786, as Johann von Charpentier, having been born in the German town of Freiberg. He studied at the Freiberg Mining Academy, and worked in the mines of Silesia before moving to Switzerland to take up the post in Bex in 1813, where he adopted the French version of his name. De Charpentier's interest in geology extended far beyond the practical needs of his mining work, and he was already convinced that it would have been impossible to transport the huge erratic boulders seen in the valleys by water. Even so, in 1815 he dismissed Perraudin's ideas as a flight of fancy, too extravagant to be even worth examining.

Undaunted, Perraudin continued to bend the ear of anyone who would listen to his ideas, and he found a sympathetic listener in the form of Ignace Venetz, a highway and bridge engineer whose work had made him familiar with the geology of the region. Like de Charpentier, Venetz had a broad scientific interest in geology, and attended meetings of the Swiss Society of Natural Sciences. In 1816, at the annual meeting of the Society, he described how heaps of rocky debris (moraines) accumulate at the termination of glaciers today, but he stopped short of going further. In 1821, he wrote a paper in which he identified similar heaps of rock five kilometres from the present-day end of the Flesch glacier as terminal moraines deposited there when the glacier was more extensive, but he didn't publish this at the time. It was only in 1829 that Venetz felt he had enough firm evidence to present to the annual meeting of the Society the suggestion that instead of just moving up and down the mountain valleys, the glaciers had once spread in a vast ice sheet across Switzerland, the Jura and other parts of Europe. De Charpentier, who was in the audience, knew Venetz well, as a fellow amateur scientific geologist, and this time he was convinced. He encouraged Venetz to publish his 1821 paper, which appeared in print in 1833, carried out his own field studies, particularly in the Rhône valley, and prepared a paper of his own for presentation to the Society at its annual meeting in 1834, held that year in Lucerne. This time, Agassiz, back from Paris, was in the audience; but, like de Charpentier when he first encountered the Ice Age hypothesis, he was far from being convinced.

Agassiz and de Charpentier had been known to each other since young Louis had been a schoolboy, when he met de Charpentier, and the older man had followed the career of the young prodigy with interest. Although deeply embroiled in his study of fossil fish, and in spite of his respect for de Charpentier, Agassiz was so annoyed at what he saw as a nonsensical idea that he determined to make him see the error of his ways. The opportunity came in the summer of 1836, when de Charpentier invited the naturalist to stay with him in Bex, and see for himself the evidence of former glaciation. Agassiz went with the intention of persuading de Charpentier to give up these wild ideas and accept the conventional ice-rafting model; but when he saw the evidence first hand, he was forced to accept that de Charpentier and Venetz were right. With the fervour of a convert, Agassiz temporarily set aside his work on fossil fishes and gathered evidence to support his much more grandiose version of the idea, arguing that there really had been a great ice sheet, engulfing Europe from the North Pole all the way to the Mediterranean Sea.

As this suggests, Agassiz got completely carried away by the idea of a great Ice Age, and went much further than the rather startled de Charpentier can have anticipated. De Charpentier was quite happy to present his evidence to the scientific community, perhaps apply a little gentle persuasion (as he had in the case of Agassiz), and wait for the evidence to convince the doubters. But Agassiz wanted to preach the Ice Age gospel, and expected all his contemporaries to fall in line at once in the light of the evidence he had seen. By the time the next annual meeting of the Society came around, at Neuchâtel on 24 July 1837, Agassiz (by now President of the Society, though still only thirty) was ready to let rip. Settling into their seats for his Presidential Address to the meeting, the luminaries of the Swiss Society of Natural Sciences anticipated an advertised discourse on fossil fishes. But Agassiz had thrown away the script, and instead offered them a lecture on Ice Ages, which he had prepared only the day before (it was this talk that introduced the term 'Ice Age', or *Eiszeit*, suggested to Agassiz by a colleague, the botanist Karl Schimper). 'Just recently,' Agassiz announced:

Two of our colleagues [de Charpentier and Venetz] have generated through their investigations a controversy of far-reaching consequences for the present and the future. The characteristics of the place in which we meet today suggest my talking to you again of a subject which, in my opinion, may be solved by the investigation of the slopes of our Jura. I have in mind glaciers, moraines and erratic boulders.

Any audience would be gripped by such a speaker; but although they listened, this audience did not like what they heard. Expecting instant acceptance of the idea, Agassiz was astonished when the talk actually provoked a mixture of anger and disbelief. Even a field trip into the Jura mountains, which Agassiz arranged in the expectation that the doubters would be persuaded once they had seen the evidence of their own eyes, descended into farce, with opponents of the Ice Age idea regarding the scars and grooves cut into the rocks as being merely damage caused by passing carriages.

But to some people outside this circle, the idea of a great Ice Age was, almost literally, a godsend. In a telling remark during that Presidential Address, Agassiz spelled out the catastrophist nature of his idea:

There is then a complete break between the present creation and those which preceded it; if the living species of our times resemble those buried in the levels of the earth, so as to be mistaken for them, it cannot be said that they have descended in direct line of progeniture, or what is the same thing that they are identical species.

This was, after all, still twenty-two years before Charles Darwin would publish his *Origin of Species*, and most biologists were still groping in the dark when it came to explaining the relationships between living and extinct forms of life. On his travels around Europe, Agassiz had visited England in 1835, and there met leading geologists including Charles Lyell (the arch-uniformitarian) and William Buckland, by then in his fifties, and a leading proponent of catastrophism. As news of the Ice Age idea spread (and Agassiz made sure it did, by giving talks at other scientific meetings), it was hardly surprising that uniformitarians such as Lyell rejected it, but it was particularly disappointing for Agassiz that Buckland also failed to warm to the idea, in spite of a visit to Neuchâtel in 1838.

The reception accorded to the Ice Age idea only stirred Agassiz into more proselytizing. He gathered more evidence in support of his case, including making his own observations on the Aar glacier, driving wooden stakes into the ice and returning each summer to measure how far they had moved as the ice flowed downhill. 'There', he later wrote, 'I ascertained the most important fact that I now know concerning the advance of glaciers, namely, that the cabin constructed by Hugi in 1827, at the foot of the Abschwung, is now four thousand feet lower down.' And then, in 1840, he published a book, *Études sur les glaciers*, which presented the Ice Age idea in language that nobody could ignore, whether they liked the idea or not:

The development of these huge ice sheets must have led to the destruction of all organic life at the Earth's surface. The ground of Europe, previously covered with tropical vegetation and inhabited by herds of great elephants, enormous hippopotami, and gigantic carnivora became suddenly buried under a vast expanse of ice covering plains, lakes, seas and plateaus alike. The silence of death followed . . . springs dried up, streams ceased to flow, and sunrays rising over that frozen shore . . . were met only by the whistling of northern winds and the rumbling of the crevasses as they opened across the surface of that huge ocean of ice.

Phew! In a two-pronged attack, in the same year that his book was published Agassiz expounded his Ice Age idea to the annual meeting of the British Association for the Advancement of Science, held that year in Glasgow, in September. The talk provoked a negative reaction from Lyell, as must have been anticipated, but this time Buckland sat in the audience without taking sides in the ensuing debate. The reason soon became clear. After the meeting, Buckland invited Agassiz and another leading geologist, Roderick Murchison, to join him on a field trip to look for signs of glaciation in the highlands of Scotland. It was on this expedition that Buckland, already leaning in that direction, was finally converted to the Ice Age cause, and Agassiz gathered evidence which helped him to refine his ideas. When Agassiz departed for a trip to Ireland, Buckland headed for Forfarshire, to visit Lyell, who had stayed on in Scotland at the home of his father. Straightaway, Buckland took Lyell out to search for evidence of terminal moraines, and straightaway they found it. On 15 October, Buckland wrote to Agassiz:

Lyell has adopted your theory *in toto!!* On my showing him a beautiful cluster of moraines within two miles of his father's house, he immediately accepted it, as solving a host of difficulties that have all his life embarrassed him. And not these only, but similar moraines and detritus of moraines that cover half of the adjoining counties are explicable on your theory, and he has consented to my proposal that he should immediately lay them all down on a map of the county and describe them in a paper to be read the day after yours at the Geological Society.

In fact, Agassiz, Buckland and Lyell each presented a paper on the Ice Age theory to the Geological Society before the year was out, and this triple whammy finally established the theory in science, not least because of the alliance between uniformitarians and catastrophists, each agreeing that the evidence of an Ice Age was incontrovertible. With the publication of Agassiz' book just a few months earlier, 1840 can be said to be the year when the Ice Age theory came in from the cold.

Agassiz had the pleasure of seeing his ideas gain credence,

but at the cost of a friendship. De Charpentier had been quietly working on his own, much more sober, book, which appeared in 1841, but attracted little attention. He never forgave Agassiz for picking up his ball and running off with it. Agassiz himself, a failed marriage behind him, went to America on what had been intended as a working visit in 1846, but made such an impact that a Chair of Geology was established for him at Harvard in 1847, and he was persuaded to stay. Apart from finding more evidence of former glaciation in North America, and enjoying a happy second marriage, Agassiz had a profound impact on the development of American science, establishing the Museum of Comparative Zoology in Harvard, but remaining, to the end of his life in 1873, a staunch catastrophist and opponent of the theory of evolution by natural selection. By the time Agassiz died, the question was not whether there had ever been an Ice Age, but what had caused the Ice Age. And one man in particular was well on the way to establishing the basis of Ice Age rhythms.

In fact, the seeds of the astronomical theory of Ice Ages can be traced back to a book published in 1842, just two years after the publication of Agassiz' own book. Written by Joseph Adhémar, a mathematician who worked as a tutor in Paris, the ideas described in the book, *Révolutions de la mer*, were confused and largely misguided – but they were based on a key fact which stimulated further investigation by others. The key fact had, in fact, been known since the seventeenth century – it was Johann

Kepler's discovery that the orbit of the Earth around the Sun is not a circle, but an ellipse, with the Sun at one focus of the ellipse. This means that at one end of its orbit it is slightly closer to the Sun than it is when it is at the other end of its orbit, so that the amount of heat falling on each square metre of the surface of the Earth each day is slightly greater in one half of the orbit than in the other. Just how this affects the weather and climate on Earth depends on the way the Earth is tilted relative to the plane of its orbit (the plane of the ecliptic), which is what gives us the cycle of the seasons. The Earth leans over slightly (at present, by about 23.5 degrees out of the vertical), and this tilt maintains essentially the same orientation relative to the distant stars over the course of a year. So for some of the time, the Northern Hemisphere leans towards the Sun, and experiences summer (long sunny days and short dark nights), while the Southern Hemisphere is tilted away from the Sun, and experiences winter. Six months later, the situation is reversed, with long dark nights and short sunny days in the Northern Hemisphere. As it happens, the Earth is at its closest to the Sun (perihelion) today on 3 January each year, during Northern Hemisphere winter; fairly obviously, this means that it is at its furthest from the Sun (aphelion) half a year later (or earlier) at the end of June, in Northern Hemisphere summer. When the tilt of the Earth is at right angles to a line joining the Earth and the Sun (on 20 March and 22 September each year), everywhere on Earth there is an equal balance between day and night.

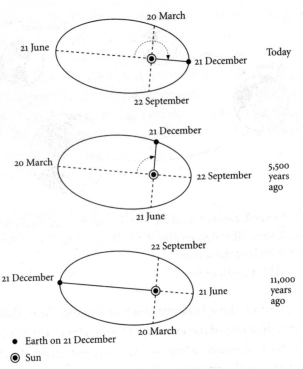

- Earth on 21 December
- Sun

Precession of the equinoxes. The dates of the equinoxes, now 20 March and 22 September, and the solstices, now 21 June and 21 December, drift slowly round the Earth's elliptical orbit with a cycle 22,000 years long. The Northern Hemisphere winter solstice today occurs when the Earth is almost at its closest to the Sun; 11,000 years ago northern winter occurred when the Earth was at the far end of its orbit.

The effect is greatly exaggerated in this illustration, based on one from Imbrie and Imbrie.

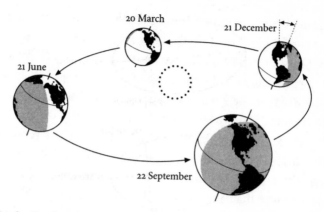

As the Earth moves round the Sun during the year, its tilt points in the same direction relative to the distant stars. This is the cause of the cycle of the seasons.
Based on a diagram from G. Kukla.

This fact alone tells you that our distance from the Sun is not the cause of the seasons; in the Northern Hemisphere, we have summer when we are furthest from the Sun. But maybe the elliptical orbit produces other, more subtle effects on climate. Several people speculated along these lines, including the astronomer John Herschel (the son of William Herschel, who discovered the planet Uranus), in the 1830s. But Adhémar came up with a fully worked out (if wrong) model of how this might cause Ice Ages – and it is particularly significant that even in the 1840s he was talking about a succession of Ice Ages, not just 'the' Ice Age. Because the Sun is at one focus of the ellipse traced by the Earth's orbit, not at the centre of the ellipse, a line passing

21

through the Sun and joining the two points in the orbit where the number of daylight hours is equal to the number of hours of darkness (the two equinoxes, on 20 March and 22 September) divides the orbit into two unequal 'halves'. The distance the Earth travels around its orbit between 22 September and 20 March is shorter than the distance it travels between 20 March and 22 September. The result is that in the Northern Hemisphere, the summer 'half' of the year (defined as the interval between equinoxes) is seven days longer than the winter half of the year, while in the Southern Hemisphere winter is seven days longer than summer. Adhémar reasoned that this explains why the Antarctic is frozen today – the Southern Hemisphere has got colder and colder over the millennia, because of the imbalance between summer and winter.*

So why should the Northern Hemisphere experience Ice Ages? This time, Adhémar went back even further in history, to the time of the Ancient Greeks. As long ago as 120 BC the astronomer Hipparchus had noticed that the direction in which north 'points' on the sky changes as time passes – he could tell this by comparing his observations of the constellations with those made by his predecessor, Timocharis, around 270 BC. By the 1840s, it was well known that this shift is caused by a stately wobble of the tilted Earth, resembling the way a child's spinning top wobbles as it spins.

---

* It's worth emphasizing that this idea is wrong, as will become clear later. Antarctica is frozen today because warm ocean currents cannot penetrate to high southern latitudes.

The wobble means that the direction in which the North Pole points traces a circle around the sky once every 26,000 years (the same is true of the South Pole, but since all the key people involved in this story lived and worked in the Northern Hemisphere, we'll describe things from their perspective). The effect is caused by the gravitational pull of the Sun and Moon, acting on the bulging equator of the not-quite-spherical Earth, and it means that the familiar Pole Star is only temporarily a good guide to the direction North; six thousand years ago, the appropriate star to use to find north would have been the one right at the end of the 'handle' of the constellation known as the Big Dipper.

In terms of the seasons, the effect of this wobble is to make the dates on which the equinoxes occur shift slowly, in step, around the Earth's orbit. The effect is slightly counter-balanced by another effect, in which the whole orbit of the Earth around the Sun is shifting around the Sun, so that each ellipse traced by the orbit is rotated slightly, pivoting on the focus occupied by the Sun, compared with the previous ellipse (the effect is caused by the gravitational pull of the other planets). Overall, the effect is called the precession of the equinoxes, and it takes roughly 22,000 years to complete one cycle. So 11,000 years ago, winter in the *Northern* Hemisphere was seven days longer than summer. You can see why Adhémar got excited about this. According to his model, this meant that the north would have been in the grip of an Ice Age at that time. He envisaged a pattern of alternating Ice Ages between Northern and Southern Hemispheres, with an

11,000 year (or 22,000 year to complete the cycle) rhythm. But then he went completely over the top. In a wild flight of fancy, Adhémar went on to suggest that the mass of ice accumulating over Antarctica would get so big that its gravitational pull would drain water from the Northern Hemisphere into the south (this really is crazy, since the gravitational pull of the ice is tiny and can be easily calculated from Newton's law of gravity). Then, when the pattern of the seasons changed and the Southern Hemisphere began to warm up, the warm water would eat away the base of the huge ice sheet, undermining it until it collapsed into the sea, creating a huge tidal wave rushing northwards. It was this aspect of his model which gave Adhémar the title of his book, so you can see how important it was to him, and how his whole model was cast in a catastrophist mould. Crazy though this particular idea was, though, Adhémar did set some people thinking about changes in the Earth's orbit as a cause of Ice Ages, and he made the crucial point that any such influences are cyclical, producing a repeating rhythm of Ice Ages. But there was another fatal flaw with his model, which was pointed out by the German Alexander von Humboldt, in 1852. The heat balance of each hemisphere of the Earth doesn't depend on how many hours of darkness and daylight there are, but on how much heat is absorbed by each square metre of the surface during those hours of daylight, over the course of an entire year. When all the factors are taken into account, including the changing distance of the Earth from the Sun, it turns out that the

Precession of the Earth. Because of the way the Sun and Moon tug on the equatorial bulge of our planet, a line joining North and South poles (the axis of rotation) traces out a cone. At present, the angle by which the Earth is tilted out of the vertical from a line joining us to the Sun is 23.5 degrees; this angle also changes, nodding up and down over a range of about 3 degrees. This nutation affects the intensity of the difference between the seasons.
After Imbrie and Imbrie

total amount of heat falling on each hemisphere over an entire year is the same as the total amount of heat falling on the other hemisphere in the year. So neither hemisphere should be getting warmer or cooler than the other because of any difference in the total amount of heat they receive from the Sun, because there is no such difference.

The person who developed the line of thought pioneered by Adhémar into a coherent model, free from any extravagant flights of fancy and with a neat explanation for how seasonal influences really could affect climate (but still wrong in detail, for understandable reasons given the state of knowledge in the second half of the nineteenth century) was the Scot James Croll. Croll was born on 2 January 1821, in the parish of Little Whitefield, a village of some ten houses at Cargill, on the banks of the River Tay, in Scotland. He was one of four brothers, but two of them died before reaching adulthood. His family farmed a small piece of land – or rather, James' mother tried to extract whatever could be wrung from the land, while his father worked as an itinerant stonemason, travelling from job to job to provide for a family that he rarely saw. Croll's father (whose name was David, and spelled his surname Croil) originally farmed a few tens of acres of land, but when young James was three the owner of the village, Lord Willoughby, demolished the houses, put all 150 acres of the land previously occupied by four tenants into one farm, and provided each of the unlucky tenants, including David Croil, with a tiny piece of land on which to build a house and live. James' formal education, such as it was, came to an end when he was thirteen, and left school to work on the couple of acres of ground where the family kept a cow; but he became a voracious reader, fascinated by books on philosophy and theology, and soon moving on to science and technology. He later wrote of his first encounter with science that 'soon the beauty and simplicity of the conceptions filled me with delight and astonishment, and I began then in earnest to study the matter'. When Croll said 'in earnest', he meant 'in earnest':

In order to understand a given law, I was generally obliged to make myself acquainted with the preceding law or condition on which it depended. I remember well that, before I could make headway in physical astronomy ... I had to go back and study the laws of motion and the fundamental principles of mechanics. In like manner I studied pneumatics, hydrostatics, light, heat, electricity and magnetism. I obtained assistance from no one.

Over the next three years, Croll developed what he described as 'a pretty tolerable knowledge of the general principles' of all these subjects. But in 1837, at the age of sixteen, he had to find a way to earn his own living. 'The bent of my mind at the time was to obtain a university education, which might enable me to follow out physical science. This, however, was a wish that could not be realized, as my father was by far too poor.' Because of his interest in theoretical mechanics, he decided to try his hand as a millwright, but soon found that he had made a mistake. He understood the theoretical principles of his new trade well enough, served out his apprenticeship and plied the trade until he was nearly twenty-two; but discovered that 'the strong natural tendency of my mind towards abstract thinking somehow unsuited me for the practical details of daily work.' That remark, which resonates strongly with one of the authors of the present book, could be the motto which sums up the rest of Croll's working life. At the age of twenty-one, Croll gave up the hopeless struggle to turn himself into a millwright, and returned to the family home (we can only guess the response of his parents) to concentrate on studying algebra, while trying to earn a little income as a carpenter. Curiously, he found that this activity quite suited him, and he might have been happy to remain a carpenter with an interest in science had it not been for a damaged elbow. He had hurt his left arm in a boyhood accident, at about the age of eleven, when a boil on the elbow was accidentally banged on a door. The old injury had never properly healed, and from time to time the elbow became inflamed and sore. Now, the repetitive movements involved in carpentry aggravated the old injury, causing severe pain and then stiffening the joint. Forced to think again about earning his keep, the ever-optimistic Croll tried his hand at working as a tea merchant (not serving tea and buns in a kind of café, as some accounts suggest, but as a retailer of leaf tea). He did well enough at this to decide to open his own shop, and well enough socially (in spite of chronic shyness) to marry Isabelle MacDonald. The couple settled in Elgin, and might have stayed there if the elbow injury had not flared up, causing serious inflammation and completely destroying and ossifying the joint. During his illness, the tea retailing business suffered, and afterwards it proved impossible to get it back into profit, partly because of this setback but also because 'the strong natural tendency of [Croll's] mind towards abstract thinking' had made him a poor businessman, who spent more time studying

philosophy and physics than he did on his accounts. Looking on the bright side, however, Croll noted that although
his elbow was now stiff, the inflammation never returned,
and 'I afterwards enjoyed better health.'

By the summer of 1850, Croll had to sell up. After a brief
spell as a salesman of electrical goods, he decided to run a
temperance hotel (he had strong religious convictions and
was a pledged abstainer), and in spite of his stiff elbow
made most of the furniture himself. The hotel opened in
Blairgowrie in 1852, and closed in 1853. The teetotal Croll
had opened the only hotel which didn't serve whisky in a
town which boasted only 3,500 inhabitants but sixteen other
hostelries all serving alcohol. There now followed the worst
time of Croll's life – four years on the road as an insurance
salesman. 'To one such as me, naturally so fond of retire-
ment and even of solitude, it was painful to be constantly
obliged to make up to strangers.' Stubbornly, Croll per-
severed with the work until his wife became seriously ill,
and they had to move to Glasgow, where she could be
looked after by her sisters (she spent more than a year
in bed). Croll took advantage of the resulting period of
unemployment ('I was now at perfect leisure') to write a
book, *The Philosophy of Theism*. In spite of his shyness, he
travelled to London with the manuscript, found a publisher,
and saw the book not only gain favourable reviews, but
actually make a small profit, although only 500 copies were
printed. In 1858, Croll got a job with a weekly temperance
newspaper devoted to campaigning for social reform, the

Photograph of James Croll. (From James Irons, *Autobiographical Sketch of James Croll* (Stanford, London, 1896).)

Glasgow *Commonwealth*. And then, in 1859, came Croll's big break. He got a job as a janitor.

To put the date in some sort of scientific perspective, 1859 was, of course, the year that Charles Darwin published his *Origin of Species*, and public interest in science was probably as high then as it is today. But even Croll could not have imagined that the janitoring job would lead to him mingling with the likes of Darwin and Lyell, and becoming a Fellow of the Royal Society. He took the job, at the Andersonian College and Museum, in Glasgow, because it gave him the solitude he craved, as well as access to a first-class scientific library. One wonders how good a janitor Croll was; he tells us that 'my duties were regular and steady, requiring little mental labour; and as my brother was staying with me, he gave me a great deal of assistance, which consequently allowed me a good deal of spare time for study.' The environment suited him down to the ground. 'I have never been in any place so congenial to me as that institution,' he wrote. 'My salary was small, it is true, little more than sufficient to enable me to subsist; but this was compensated by advantages for me of another kind.' He meant the library, and the peace and quiet, allowing him to give rein to his 'strong and almost irresistible propensity towards study'. At first, Croll studied physics, and in 1861, when he was forty, published a paper on electricity. This conjures up a delicious image of the editors of the learned journal concerned (and, soon, the editors of other learned journals) receiving a paper from the highly respectable

Andersonian, but being totally unaware that it came from the janitor. In the early 1860s, however, 'the' Ice Age was once again a topic of lively discussion among geologists, not least because of its relevance to the debate about evolution stirred by the publication of Darwin's book. It was only a matter of time before Croll turned his attention in that direction, and he began reading up on the subject in 1864. He soon came across Adhémar's book, now nearly a quarter of a century old, and although he was aware of the flaws with the detailed model of Ice Ages proposed by Adhémar, he became convinced that there must be some other way in which the astronomical rhythms really could act to cause Ice Ages. And Croll had one key piece of astronomical information which had been unknown to Adhémar (or at least, not considered by him) in 1842. As well as the effect of the precession of the equinoxes, on which Adhémar's model had been built, the shape of the Earth's orbit is always changing, from slightly more extended (more elliptical) to more precisely round (more circular) and back again, with a repeating rhythm roughly 100,000 years long. In fact, the evidence for this effect, known as a change in the orbital eccentricity of the Earth, had been provided at almost the same time that Adhémar completed his book, by the French astronomer Urbain Leverrier.

Leverrier is best known today for his successful prediction, based on Newton's law of gravity and changes in the orbits of the known planets, of the disturbing presence of Neptune, which he made in 1846 and which led to the discovery of Neptune, exactly where Leverrier had predicted it must be (essentially the same prediction was made a little earlier, but not widely publicized, by the British astronomer John Couch Adams). But Leverrier had first cut his astronomical teeth a few years earlier, with calculations of the changing orbital eccentricity not just of the Earth, but also Mercury, Venus and Mars; all these changes are caused by the gravitational influences of the planets on one another, and it was from this work that Leverrier went on to study the orbits of Jupiter, Saturn and Uranus, and the orbits of comets, soon leading to his prediction of an 'extra' planet. All of this work required painstaking and laborious calculations, in those days when everything had to be done literally by hand, using pencil (or pen) and paper. Leverrier's work on the inner planets (the work relevant to the story of Ice Ages) actually provided details of the orbit of the Earth at intervals of 10,000 years, for the 200,000 years centred on AD 1800 – enough for Croll's first attempt at finding a relationship between the astronomical influences and climate. The change in eccentricity is measured in terms of the distance between the two foci of the ellipse, as a percentage of the long axis of the ellipse. For a perfect circle, the two foci merge to become one, with no distance between them, so the eccentricity is zero. Today, the eccentricity of the Earth's orbit is about 1 per cent, but Leverrier showed that at its most extreme the Earth's orbit has an eccentricity of roughly 6 per cent. Because Leverrier's calculations showed that the Earth's orbit was in a more highly eccentric state 100,000 years ago, while for the past 10,000 years or so it has been in a low eccentricity state, and since the world is warmer now than it was in the past, Croll speculated that some effect associated with high eccentricity must be responsible for Ice Ages. This was no more than a speculation when Croll published his first paper on the subject in 1864, but the paper provoked so much interest that he resolved to look into the problem more fully. 'Little did I suspect', he recalled, 'that it would become a path so entangled that fully twenty years would elapse before I could get out of it.' Happily, we do not have to follow all the twists and turns and false trails of that tangled path but can cut straight to the definitive version of Croll's model, as presented in his book *Climate and Time*, published in 1875, when its author was fifty-four years old (the 'twenty years' was a bit of an exaggeration; Croll's later work on Ice Ages added little to the model as presented in his great book). By then, Croll had long been a full-time, professional scientist – in 1867 he took a post in Edinburgh with the Geological Survey of Scotland (essentially a clerical job, with the Director of the Scottish Geological survey, Archibald Geikie, encouraging Croll to carry out his own research) and just a year after his book on Ice Ages was published, in 1876 he was elected as a Fellow of the Royal Society, a few months after receiving the honorary degree of LL.D. from the University of St Andrews. He was forced to retire early, in 1880 at the age of fifty-nine, through ill health; he had long suffered from severe headaches, and now developed a heart problem as well. Croll lived quietly with his wife near Perth for the next ten years, still working on his Ice Age model and then on another philosophical book, in spite of the almost constant pain caused by his headaches. He died there on 15 December 1890, soon after the publication of that book, *The Philosophical Basis of Evolution*.

One fruit of Leverrier's painstaking pages of calculations had been his derivation of a relatively simple formula which makes it possible to calculate the eccentricity of the Earth's orbit at different times in the past (or, indeed, in the future). As Croll put it, this was sufficient 'for ordinary astronomical purposes, but far too limited to afford information in regard to geological epochs'. In his extended work on Ice Ages, Croll used the formula derived by Leverrier to extend the eccentricity calculations back three million years into the past and forward one million years into the future, calculating the details of the Earth's orbit at intervals 50,000 years apart. Unlike any of his predecessors, he plotted the results on a graph, where the pattern of highs and lows could be easily picked out. He found that during that time there have been widely spaced intervals when the eccentricity was high, for ten or twenty thousand years, separated by much longer intervals, roughly a hundred thousand years long, with low eccentricity.

Leverrier's calculations had already shown that whatever kind of orbit the Earth is in at any particular epoch, the amount of heat received from the Sun over the course of an entire year stays the same; but Croll followed up the idea that it might be the way the heat is distributed between the seasons which matters, since this is undoubtedly affected by the eccentricity. When the orbital eccentricity is low, and the orbit is circular, the amount of heat received by the whole planet from the Sun each week is the same throughout the year; but when the orbit is more elliptical, with high eccentricity, the Earth receives more heat in a week at one end of its orbit, closest to the Sun, and correspondingly less heat in a week at the other end of its orbit, farthest from the Sun. Depending on which hemisphere you live in, this may mean that when the orbit is more eccentric there is more difference between the seasons, with cold winters when the Earth is farthest from the Sun and hot summers when it is closest to the Sun; or (in the other hemisphere) it may mean that the eccentricity effect smooths out the difference between the seasons, keeping summers cool and winters mild. Croll argued that what was needed to build an Ice Age was a series of very cold winters, so that there would be more snowfall, building up white snowfields and ice sheets which would reflect away the summer heat from the Sun to keep the hemisphere cool. He was one of the first scientists to develop the idea of feedback in any context, and although it happens that he got the detail of this influence backwards, his model would be important historically for that reason alone. 'Underground temperature', wrote Croll, 'can only affect climate through the surface. If the surface could, for example, be kept covered with perpetual snow, we should have a cold and sterile climate, although the temperature of the ground under the snow was actually at boiling point.' The other link in Croll's chain of reasoning was that the precession of the equinoxes must come into the story as well, since that is what determines in which part of the orbit winter occurs. Croll came to the same conclusion as Adhémar (but for a different reason) – that the Earth should experience an Ice Age every 11,000 years or so, first in one hemisphere then in the other. But he did not suggest that there had been an Ice Age 11,000 years before the present day, because for the past 80,000 years or so the Earth's orbit has been in a state of low eccentricity. According to Croll, that should have been a long Interglacial, following an Ice Age which peaked around 100,000 years ago, during the last interval of extreme orbital eccentricity. He predicted that the precession effect would be found to be a strong influence on climate only during those intervals when the orbital eccentricity was high, and thought that under those conditions there would be an Ice Age in whichever hemisphere experienced winter while at the farthest end of the Earth's orbit from the Sun.

But he didn't stop there. Worried that the rather small changes in the balance of heat between summer and winter caused by the orbital variations might not be enough to drive the rhythm of Ice Ages, Croll looked for a major influence on climate which could be triggered by a relatively small change in the balance of the seasons. He was one of the first people to appreciate the major influence of the great ocean currents on climate, and was the first person to work out the link between the trade winds (essentially driven by convection in the atmosphere stirred up by the Sun heating the surface of the Earth) and the flow of these currents, pushed by the winds. He reasoned that the change in the balance of heat between the hemispheres when one polar region cooled would increase the strength of the trade winds, blowing from the hotter part of the world to the colder region in an attempt to even out the temperature, and also change their direction somewhat. This would change the pattern of the ocean currents. In particular, he noted that a relatively small shift in the westward flowing current of the equatorial Atlantic Ocean could make it flow either northward past the bulge of Brazil and up past North America, or southward past the bulge of Brazil, and down past South America. The potential climatic consequences of such a shift, which could be triggered by a relatively small outside influence, are clear from the Prologue. Once again, Croll was at the forefront of thinking about feedbacks, and the way in which they can magnify small initial disturbances. He also appreciated the broad nature of the looping 'river' of global ocean currents, with a constant flow, under present day conditions, of warm surface water from the Southern Hemisphere to the north and what he described as cold 'under currents of equal magnitude' from the Northern Hemisphere to the south. Finally, in *Climate and Time* he pointed out that there is yet a third astronomical influence on climate, which really ought to be taken into account. The tilt of the Earth's axis (the amount it leans out of the vertical) also varies as time passes, as the great French mathematician and astronomer Pierre Simon de Laplace had shown at the beginning of the nineteenth century (Leverrier also calculated how the tilt changes). It can be as little as 22 degrees, or as much as 25 degrees (as we mentioned, it is now about 23.5 degrees). This affects the intensity of the seasons. Neither Leverrier nor his predecessors had worked out exactly how the angle of tilt changes as time passes, and nor did Croll; but he guessed that Ice Ages would be more likely when there was less tilt (with the Earth more upright) because that would minimize the amount of heat received by the poles during the year (to be fair, it was rather more than a guess, with calculations of how much heat is received by the poles for different degrees of tilt, but something less than a fully worked out model). Once again, Croll was ahead of his time; but by 1875, with the publication of *Climate and Time*, all the pieces of the puzzle of the astronomical influences on climate had been identified, and all that remained was to work out exactly how they should fit together.

Croll's model initially met with a warm response. Improving geological techniques showed that there had indeed been a repeating succession of Ice Ages, separated by Interglacials, just as the model required. But with the state of the art at the time, this could be no more than circumstantial evidence in support of the astronomical model. There was simply no way to date geological strata accurately enough to tell if there had been alternating

Northern Hemisphere and Southern Hemisphere Ice Ages, fluctuating with the rhythm that Croll proposed. But even without this level of sensitivity, by the beginning of the twentieth century there was a mounting weight of evidence that the latest Ice Age had ended not around 80,000 years ago, as Croll had proposed, but around 10,000 years ago. Between 100,000 years ago and 80,000 years ago, when according to Croll's model the world should have been warming out of an Ice Age, it was in fact plunging into an Ice Age. Clearly, the model was wrong. But none of the Victorians realized that the nature of its wrongness – the fact that the climate was actually changing exactly the opposite way from what the model predicted – was an important clue to how the astronomical influences really do work. The astronomical model of Ice Ages fell from favour, only to be revived when a young Serbian engineer developed a passion for the calculations required to combine all three of the astronomical influences into a single description of how the amount of heat arriving at different latitudes of the Earth at different times of year has varied over the millennia.

Two

The Serbian's Ice Age

Milutin Milankovitch was born in the small Serbian town of Dalj, on the banks of the Danube, on 28 May 1879 (the same year that Albert Einstein was born). At that time, the entire Balkan region was an unstable buffer zone between two decaying empires, the Austro-Hungarian Empire to the north and the Turkish (Ottoman) Empire to the south. Serbia could trace its existence as a nation back to the seventh century, but it had been overrun by the Ottomans in 1389, and came under long-term Turkish rule in 1459. Various wars changed the boundaries in the region down the centuries; after 1829 Serbia became an autonomous principality under Turkish suzerainty but with Russian protection, and after the Russo-Turkish War of 1877–8 (when the Serbs fought on the side of the Russians) it became independent, being proclaimed a kingdom in 1882. So at the time Milankovitch was born this part of the region was involved in major upheavals, and as he was growing up there was a more or less permanent threat from Austria-Hungary, which annexed Bosnia and Herzegovina in 1908.

The political upheavals continued, and would have a major influence on his life, affecting his opportunities to work on the astronomical model of Ice Ages, but not in the way you might have expected.

Milankovitch trained as a civil engineer at the Institute of Technology in Vienna, where he received his Ph.D. in 1904. He worked in Austria for five years on various projects as an engineer, developing a reputation as an expert in the design of large concrete structures. But when he was offered the post of Professor of Applied Mathematics at the University of Belgrade, he was happy to accept, even though it meant leaving sophisticated Vienna for an academic backwater. The move was prompted partly by a desire to return home, partly by the knowledge that Serbia (where nationalist patriotic feelings ran high in the newly independent country) needed trained engineers, and partly by a feeling that he wanted to achieve something in science, solving some big (he later used the word 'cosmic') problem. In his own words, he was 'under the spell of infinity'. Milankovitch was based in Belgrade for the rest of his working life, and within two years he had found the problem that would occupy him for the next thirty years – but always, strictly speaking as a hobby, worked on at home, alongside his day job as a teacher and engineer. According to Milankovitch's reminiscences (which should perhaps be taken with a pinch of salt) the inspiration for his life's work came in 1911 over a bottle of wine with a friend, celebrating the publication of the friend's first book of poetry. Under the influence of the wine and the euphoria of his first success, the young poet resolved to write an epic novel; Milankovitch, riding on a similar flight of fancy, declared that he would devise a mathematical model which would describe the climate not only of the Earth but of Mars and Venus as well, not just for today but for any time in the past or in the future. And unlike many such flights of fancy, Milankovitch not only remembered his but carried it through (we are not told if the poet succeeded in writing his novel). Whatever the source of his inspiration (and he must surely have been thinking about the climate problem before that bottle of wine was ever uncorked), Milankovitch's colleagues at the University of Belgrade were convinced that he had to be either drunk or crazy to contemplate such an idea. After all, we know from direct experience what the climate on Earth is like. Why bother to calculate it from first principles?

To Milankovitch, the real reason was the same as the reason George Leigh Mallory gave when he was asked why he wanted to climb Mount Everest: 'Because it's there.' But for public consumption Milankovitch pointed out that a successful model of the kind he hoped to derive would make it possible to investigate the temperature on Earth in places where nobody had ever lived, let alone made meteorological observations, such as the polar regions and the broad expanse of the oceans, as well as probing the climates of other planets in the Solar System:

Milutin Milankovitch in a portrait painted by Paja Jovanovic in 1943. (Courtesy of Vasko Milankovitch.)

For the same oven, the sun, that supplies our Earth with heat also heats those planets that are covered with solid crusts. Therefore the results of the new theory would also apply to these planets. It could give us the first reliable data about the climate of these distant worlds.

This was actually a rather topical point, in the second decade of the twentieth century. At the end of the nineteenth century, the Italian astronomer Giovanni Schiaparelli had discovered what he described as *canelli*, meaning 'channels', on the surface of Mars; the mistranslation of this word as 'canals' provoked a wave of speculation about the possibility of life on Mars, and inspired the American Percival Lowell to set up an observatory at Flagstaff, Arizona, dedicated to the search for life on Mars (it also inspired H. G. Wells to write *The War of the Worlds*, first published in 1898). A way to calculate what the climate on Mars was like, and whether or not it could support running water, was of genuine scientific interest at the time Milankovitch started his work.

What Milankovitch proposed went far beyond the work of the pioneers Adhémar and Croll. They had sketched the outlines of an astronomical model of climate change, but Milankovitch wanted to calculate every detail, in principle for every latitude on Earth (and on the other planets). It was a Herculean task. But he started with one great advantage. In 1904, the German mathematician Ludwig Pilgrim had published detailed calculations of the way all three of the orbital parameters of the Earth (precession, eccentricity, and tilt) had varied over the past million years. 'All' that Milankovitch had to do was use this information to calculate how, as the millennia passed, the combined influence of these three effects at work would change the amount of heat arriving at each square metre of the surface of the Earth, at any chosen latitude, in any season; and, remember, with all the calculations being carried out using pen and paper. It required a certain kind of mind – methodical, painstaking, patient and unruffled by setbacks (such as the occasional war). Milankovitch reckoned that he started out on the task, when he was thirty-two years old, at exactly the right time:

Had I been somewhat younger I would not have possessed the necessary knowledge and experience . . . Had I been older I would not have had enough of that self-confidence that only youth can offer.

He worked on the calculations mostly in his study at home, virtually every day, and also took the work with him on holidays. The routine stayed much the same for the next thirty years, and was described by his son, Vasko Milankovitch, in a talk given to an international symposium on the Milankovitch Model (as it became known) held at the Lamont-Doherty Geological Observatory at Palisades, New York, in the first week of December 1982:

When father did not have to go to the university for his lectures, the two [Milankovitch and his dog, Teddy] would retire to his study after breakfast. Father would smoke a pipe and work, while

Teddy lay beneath the leather divan. The pipe generally lasted an hour. Once he finished smoking, father could not stand the smell of the tobacco ... the window would be flung open. The dog would jump on the window sill (he, too, probably did not like the smell) and father would leave the room ... ten minutes later, he would close the window and return to work at his desk, with the dog once again under the divan.

Coffee (black) was served by Milankovitch's wife promptly at ten o'clock, and occupied just ten minutes before it was back to work. Lunch at one was followed by a short siesta and a cigar, then more calculations until six, when work ceased for the day. A stroll before dinner, which was a leisurely meal taken at eight, where the family discussed the topics of the day, was followed by an early night, with bed at ten providing time to read (never anything related to work) for an hour before settling down, usually to think for a considerable time before sleeping.

I was always intrigued by his method of work – never rushing nor delegating anything, including even drawing and translating into French or German. He always made thorough preparations by outlining relevant points before precisely detailing the body of the article, then rewriting prior to typing the final copy himself. A simple reply to any correspondence would be treated in the same personal manner.

You might think that this, and his lecturing duties, left Milankovitch with little time for anything else. In fact, he was widely in demand as a consultant for civil engineering work, and supervised all the large projects using reinforced concrete built for the Jugoslav Navy and Air Force in the years following the First World War and the establishment of that country out of a patchwork of Balkan states. He also found time to advise the Sultan of Tunisia on climate in connection with an agricultural project, receiving in return an elaborate decoration which he gave to his son to play with, and wrote several books on the history of science and a popular book about astronomy. Not bad for someone who, in his own words, 'cooked on a low fire'.

In spite if his methodical approach, the work for which Milankovitch is best remembered got off to a sticky start, and progress was slow as he wrestled with the recalcitrant equations. The first breakthrough came in circumstances which might not seem to be conducive to the solution of a scientific puzzle, in the midst of a conflict known as the First Balkan War, which broke out in 1912. Serving on the staff of the Serbian army, Milankovitch was with the troops that crossed into the Turkish Empire and inflicted a rapid defeat on the enemy; it was while watching the fighting that he suddenly had a flash of insight which showed him the way around the mathematical logjam that had been holding him up. It was a classic example of the way the answer to a problem you have been struggling with can pop into your head once you stop looking for the solution. Back in Belgrade, Milankovitch made rapid progress and published several papers on the problem of the astronomical model of Ice Ages in 1912, 1913 and 1914. Even at this early stage of the project, he had done enough to show that the changes in orbital eccentricity and the precession of the equinoxes can produce effects on the balance of heat at different latitudes big enough to influence the size of the ice sheets on Earth, and he also investigated the effect of changes in tilt. But his major results were published in Serbian, and no mainstream astronomers or geologists noticed this work, carried out by a civil engineer and written in a language few (if any) of them read. Having laid the foundations, Milankovitch now needed time to complete his detailed calculations. With his other duties at the University of Belgrade, and as a consulting engineer, the prospect of completing this task must have seemed remote indeed; but in 1914 the unstable political situation in the Balkans precipitated the First World War, and incidentally gave Milankovitch an almost ideal opportunity to carry out this work.

This is not the place to go into the horribly complicated political background to the First World War in any great detail, but in 1914 Europe was divided into armed alliances, each held together by a web of complicated treaties and by mutual distrust of the other armed alliances. The immediate trigger for the war (as every schoolboy used to know, but doesn't any more) was the assassination of the heir to the Hapsburg throne, Archduke Franz Ferdinand, in Sarajevo, by a Serbian extremist. Austria-Hungary declared war on Serbia, Russia mobilized her army in support of Serbia,

Germany (feeling threatened by the Russian mobilization) declared war on both Russia and Russia's ally France, then invaded Belgium, which brought Britain into the war on the side of the French and Russians (and, of course, the Serbians). Japan joined in as an ally of Britain. As clauses in various treaties were triggered by all this activity, Italy joined France, Britain and their Allies, while Turkey and Bulgaria joined the German-speaking countries. The images of trench warfare on the Western Front, the collapse of Russia and the rise of communism to the east, and Ernest Hemingway's account of war on the Italian front in *A Farewell to Arms* have all helped to give the impression to many in the English-speaking world that the war in the Balkans was something of a backwater. This is true up to a point, but the suffering there was immense, and Serbia itself lost a quarter of its population in the conflict. All of this, though, passed Milankovitch by. He was visiting his home town of Dalj when war broke out at the end of July 1914, and was promptly captured by the Austro-Hungarian army. As a reservist with the Serbian army, he was imprisoned at Esseg, an experience which he later described in graphic terms:

The heavy iron door was closed behind me. The massive rusty lock gave a rumbling moan when the key was turned . . . I adjusted to my new situation by switching off my brain and staring apathetically into the air. After a while I happened to glance at my suitcase . . . My brain began to function again. I jumped up, and opened the suitcase . . . In it I had stored the papers on my cosmic problem . . . I leafed through the writings . . . pulled my faithful fountain pen out of my pocket, and began to write and calculate . . . As I looked around my room after midnight, I needed some time before I realised where I was. The little room seemed like the nightquarters on my trip through the universe.*

But Milankovitch's incarceration lasted only a few months. Following representations made by a leading Hungarian academic, in the interests of science Milankovitch was paroled just before Christmas, on condition that he carried out only peaceful activities, lived in Budapest and reported to the police once a week. For most of the next four years, Milankovitch was able to work on his 'cosmic problem' uninterrupted, in the library of the Hungarian Academy of Sciences. It took him two years, working, as he put it, 'without hurry, carefully planning each step', to come up with a complete mathematical model describing the climate on Earth today, and the next two years to adapt this model to describe the present-day climates of Mars and Venus. When he returned home at the end of the war, Milankovitch wrote all this work up, together with his earlier conclusions, and published it (this time, in French) in a book in 1920. Even though the language was more accessible to a wider number of people, the book did not immediately strike a chord with geologists – with one key exception.

* Translations of Milankovitch's writings taken from Imbrie and Imbrie.

Wladimir Köppen was a Russian-born German meteorologist, who by 1920 was one of the leading lights in the
young science of climatology. He had been born in
St Petersburg in 1846, and was educated first at the University there and then in Heidelberg and Leipzig. He began his
career with the Russian meteorological service, in 1872,
but soon moved to Germany, where he became, in 1875,
Director of the Meteorological Research Department of the
German Naval Observatory at Hamburg. Köppen held this
post until he retired at the end of the First World War.
Starting in 1900, over a period of several decades Köppen
developed the classification system for climate types, represented by letters of the alphabet, which is still widely used
on maps today. His contribution to the understanding of
Ice Ages was made in collaboration with the geologist Alfred
Wegener (the pioneer of the idea of continental drift), who
was also Köppen's son-in-law. Wegener had been born in
Berlin in 1880, and studied in Heidelberg, Innsbruck and
back in Berlin, completing his Ph.D. in astronomy in 1905.
But he had a longstanding interest in meteorology, and
in 1908, after working as a meteorologist on a field trip
to Greenland, he was appointed to a joint lectureship in
astronomy and meteorology at the University of Marburg.
It was there that he began to develop his ideas on continental drift, which were more fully worked out during his
convalescence after being wounded in the fighting in the
early months of the First World War; after recovering from
his wounds, Wegener spent the rest of the war with the

meteorological service of the German army. At the end of the war he moved to Hamburg, and then, in 1924, he became Professor of Meteorology and Geophysics at the University of Graz, in Austria. It was in the same year that Köppen and Wegener published their joint book *Die Klimate der Geologischen Vorzeit (The Climates of Geological Prehistory)*, which included a discussion of the Milankovitch Model, including Köppen's key contribution to the development of these ideas, as well as relating climates of the past to continental drift.

The first inkling Milankovitch had that anybody had noticed his book of 1920 was when a postcard arrived from Hamburg, where the 76-year-old climatologist had immediately recognized the importance of Milankovitch's work. That postcard led to a lengthy correspondence between the two men; the flow of information between Milankovitch and Köppen and Wegener helped both sides, but as far as the understanding of Ice Ages is concerned the single most important thing to emerge from these discussions was Köppen's realization of the key season in the Ice Age saga. Adhémar and Croll had thought that the decisive factor in encouraging ice to spread across the Northern Hemisphere must be the occurrence of extremely cold winters, resulting in increased snowfall. At first, Milankovitch had shared this view. But it was Köppen who pointed out that it is always cold enough for snow to fall in the Arctic in winter, even today, and that the reason that the Northern Hemisphere is not in the grip of a full Ice Age at present is because the 'extra' snow melts away again in summer. He reasoned that the way to encourage the ice to spread would be to have a reduction in *summer* warmth, because then less of the winter snowfall would melt. If less snow melted in summer than fell in winter, the ice sheets would grow – and once they had started to grow, the feedback effect of the way the ice and snow reflect away incoming solar energy would enhance the process. This exactly turned Croll's model on its head. Instead of predicting that the Northern Hemisphere should have been warming into a long-lasting Interglacial about 80,000 years ago, the astronomical model actually suggested that the Northern Hemisphere should have been plunging into an interval of particular cold at that time, once the importance of cool summers was appreciated. Instead of saying that the Northern Hemisphere cooled around 10,000 years ago, the model now said that it should have been warming out of the Ice Age and into an Interglacial around that time.

Armed with his faithful fountain pen, Milankovitch set to on another round of calculations, working out how the three astronomical influences had affected the amount of summer heating at latitudes 55, 60 and 65 degrees north over the past 650,000 years. 'I did my calculations for a full one hundred days from morning until night and then presented the results graphically by drawing three notched, curved lines to illustrate the changes in summertime radiation.' The deep notches that appeared on Milankovitch's curves corresponded with the times in the recent geological past when Northern Hemisphere summers had been particularly cold; and Köppen soon confirmed that the timing of these notches seemed to match up with the known geological dates of the advance of the Alpine glaciers. It was the inclusion of these notched curves in the book by Köppen and Wegener that first drew the Milankovitch Model to the attention of a wide audience.

As Milankovitch continued his calculations, working out data points for intervals closer together in the geological past, his graphs became less notchy and more rounded, but still showed the same pattern of changes. He worked out in detail the three periodic variations in climate associated with the astronomical influences and identified the geographical regions where they had the greatest influence – the roughly 100,000 year long rhythm associated with the changing ellipticity of the Earth's orbit, the influence with a rhythm some 41,000 years long associated with the changing tilt (particularly important at high latitudes), and the rhythm about 22,000 years long associated with the precession cycle (less important at the poles, but more important at low latitudes). The combination of all three rhythms produces an ever-changing pattern of seasonal heating of different regions of the globe, which Milankovitch could then use to calculate how the ice sheets would advance and retreat in response. In this final phase of calculation, he used actual observations of the height above sea level at which snow lies all year long for mountains in different parts of the world. At the poles, the snowline is essentially at sea level; near the equator, where there is stronger solar heating, it occurs at high altitudes. Using these observations Milankovitch was able to work out an empirical rule relating snowcover to solar heating, and use this to relate the advance and retreat of the glaciers to the changes in seasonal heating predicted by his astronomical model. These results were published in 1938, around the time that Milankovitch began writing up a full account of his lifetime's work on the Ice Age problem. Usually referred to by its English title, *Canon of Insolation and the Ice Age Problem*, the original German edition of the book was being printed in Belgrade in April 1941, when Germany invaded Jugoslavia during the Second World War. By then, Milankovitch was sixty-two and had completed his work on the 'cosmic problem' he had chosen exactly three decades earlier. He lived on until 1958, seeing his model fall from favour in the 1950s, but quietly confident that it would ultimately stand the test of time.

Before we look at the changing level of esteem in which the Milankovitch Model was held, both during Milankovitch's lifetime and afterwards, this is a good place to pause and take note of a key feature of the model, which was never expressed in quite this way while Milankovitch was alive, but which strikes to the heart of its importance in understanding the recent climate of the world. As we mentioned earlier, given the present day geography of our planet – the distribution of the continents and oceans – the natural state of the Earth is in a full Ice Age. Köppen was correct in highlighting the importance of summer warmth in influencing the advance and retreat of the ice in the Northern Hemisphere. But, in a sense, he, too, got the argument backwards. It isn't so much that Ice Ages occur when the astronomical influences conspire to produce particularly cool summers; rather, what matters is that Interglacials only occur when the astronomical influences conspire to produce unusually warm summers, encouraging the ice to retreat. Without all three of the astronomical rhythms working in step in this way, the Earth stays in a deep freeze. And that is why the actual pattern of climate over the past few million years has been one of long Ice Ages (in fact, a single long Ice Epoch) interrupted by short-lived Interglacials, like the one we are living in now.

This was not appreciated in the 1920s and 1930s, when it seemed that the astronomical model predicted that Ice Ages only occurred at the times corresponding to the deepest notches in Milankovitch's summer insolation graphs, with relatively long warm intervals in between them. It was because the geological record seemed to show just such a pattern over the past few hundred thousand years that the model was warmly received in some quarters during the second half of the 1920s, and gained widespread support in the 1930s and 1940s, at least among European geologists, although some found it hard to understand that although the model predicted that Ice Age summers were 6.7 °C colder than today, it also said that Ice Age winters were 0.7 °C warmer than today. The problem was, though, that the geological chronology was far from being well-determined before 1950, and in addition, in a piece of circular wishful thinking, Milankovitch's Ice Age graphs were used by many geologists to place dates on Ice Age debris found in different parts of Europe – not a bad idea in principle, but in practice it is very hard to identify which Ice Age remains belong with which of Milankovitch's notches, unless there is an independent dating technique available.

The first such independent dating technique became available to geologists at the beginning of the 1950s. This is the famous radiocarbon calendar, which is based on measurements of the amount of radioactive carbon (carbon-14) in organic remains found among the geological debris. Carbon-14 is produced in the atmosphere of the Earth by the interaction of cosmic rays with atoms (strictly speaking, the nuclei of the atoms) of nitrogen in the air. Carbon-14 is chemically identical to the common stable form of carbon, carbon-12; the two versions of the same element are known as isotopes. Once it is produced, carbon-14 can, like carbon-12, combine with oxygen to form carbon dioxide, which is absorbed by plants during photosynthesis, and becomes part of the structure not only of the plants themselves but of animals which eat the plants. All living things continue to absorb radioactive carbon in small (and completely harmless) amounts in this way as long as they are alive. All living things contain both carbon-14 and carbon-12, with the same ratio of carbon-14 to carbon-12 in everything alive on Earth at the same time.

But once those living things die, they cannot absorb any more of these radioactive atoms, and the ones that are already in their tissues are gradually lost as they are converted back into stable atoms of nitrogen by the process of radioactive decay. Each radioactive isotope decays in a similar fashion, but each has its own characteristic rate of decay, known as the half life, during which half of the atoms in a sample decay. In one half life, half the atoms decay; in the next half life, half the remainder (a quarter of the original sample) decay, and so on. In the case of radiocarbon, the half life is 5,730 years, which means that after eight half lives (about 45,000 years) only $1/_{256}$ of the original radioactive carbon is left. If you know how much carbon-14 was present to start with (or at least, the proportion of carbon-14 relative to the more common, stable form of carbon, carbon-12) then you can work out how long it is since a piece of organic material was alive by measuring the ratio of carbon-14 to carbon-12 in the remains today. The process is complicated by the fact that the amount of carbon-14 being produced in the atmosphere each year varies as the cosmic ray flux varies, but the radiocarbon calendar can be calibrated by measuring the radioactivity of samples of organic material with known ages – in particular, samples of wood from very old trees, which can be dated independently by counting the number of tree-rings they contain. The technique is very good for dating organic remains laid down in the past few millennia, and has proved a great boon to archeologists; but it is extremely difficult to push it back farther than 40,000 years into the past, corresponding to about eight half lives, because there is so little radiocarbon left to measure in older material.

When the technique was applied to debris from the latest Ice Age, in the early 1950s, it soon showed that the pattern of glacial advance and retreat was more complicated than had been thought, with what had been thought to be one set of glacial debris really being a mixture of material too old to be dated with the radiocarbon technique and younger material dated to around 18,000 years ago, suggesting a late advance of the ice followed by a rapid retreat around 10,000 years ago. Frustratingly, the technique was just not powerful enough to probe back long enough in time – 80,000 to 100,000 years – to provide a real test of the Milankovitch Model. Inevitably, though, some people tried to do this, and although the evidence was extremely dubious, it seemed to show that the movement of the edge of the ice sheet in places like Quebec over the past 80,000 years or so did not match the changes in insolation calculated by Milankovitch. Over a decade and a half, more or less centred around the year Milankovitch died, the Milankovitch Model fell from grace, and by the middle of the 1960s you would have been hard pressed to find a geologist or meteorologist who regarded the model as being anything more than an historical curiosity. With hindsight, the evidence against the model can be seen to have been poor and inconclusive, not least because like all previous attempts to unravel the complexities of past climatic change it was based on a geographically and chronologically patchy collection of data from isolated places around the world, those locations where the right kind of geological remains just happened to be available for analysis. What was needed was a way for nature to take the temperature of the entire Earth, a kind of natural global geological thermometer, and to preserve a record of those global temperature fluctuations in geological locations where it could remain undisturbed until the present day. It was only in the 1970s that it proved possible to locate and interpret just such a natural record of the changing global climate – but when this was done, the astronomical rhythms finally became established as an important component not only in producing those patterns of temperature variations, but thereby producing ourselves as well.

Three

Deep Proof

The best place to find out what the climate of the Earth was like in the past is at the bottom of the deep ocean. This is by no means a new insight – it goes back at least to the time of James Croll, who wrote:

In the deep recesses of the ocean, buried under hundreds of feet of sand, mud and gravel, lie multitudes of the plants and animals which . . . were carried down by rivers into the sea. And along with these, must lie skeletons, shell and other exuviae of the creatures which flourished in the seas of those periods.

It's the second part of that remark which is particularly relevant to our story. Different kinds of sea creatures flourish under different climates, and in particular at different ocean temperatures. These creatures include the tiny planktonic forms that float about near the surface. When they die, their shells and other debris fall to the sea bottom. If and when the climate changes, the variety of marine creatures living above these remains changes, and in due course a different variety of shells and other remains falls on top of the first set of remains. Layer by layer, the mud of the sea bed builds up, and each layer contains the remains of the creatures best suited to the climate at the time that layer was being laid down. The undisturbed mud of the sea bed, as Croll appreciated, provides a chronology of past climates, with the most recent material at the top and the older material further down. The problem would be to extract samples of this mud without disturbing its layered structure, to analyse the remains found in different layers, and (by no means least) to find a way to put absolute dates on the layers, instead of just being able to say that one was older than another. As is so often the case in science, progress had to await the development of the appropriate technology – in this case, until the second half of the twentieth century, a hundred years after Croll became interested in the puzzle of past climates.

The first studies of the sea bed were actually made in the 1870s, when a British expedition on board HMS *Challenger* circumnavigated the globe making scientific observations which included dredging up samples from the sea floor. They found that, just as Croll had surmised, the sea bed near the continents is covered with debris washed down from the land. But out in the broad oceans, the sea floor is covered with a fine mud which contains the microscopic remains of the tiny floating organisms found in the sea above. One kind of shell, the remains of a variety of plankton known as foraminifera (or just forams), was particularly common in the temperate and tropical regions, while another kind of debris, made up of the shells of the kinds of plankton known as radiolaria and diatoms, dominated in the cold waters of the Arctic and around Antarctica. Even this first expedition showed that there was climatic information contained in the ooze from the sea bed. But this was only information about the present-day climate; how could the deeper layers of ooze be probed and analysed?

The solution was to drop hollow steel pipes vertically into the sea bed, so that their weight would drive them into the mud. When the pipes were hauled back on board ship, the mud inside the pipe could be extracted as a cylindrical core, with its layered structure intact. Unfortunately, because of the resistance of the water, which stops the pipes building up any great speed as they fall, this kind of 'gravity coring' can only extract cores about a metre long – the pipes just won't penetrate any deeper into the ooze. This was better than nothing, and provided the first evidence, in the 1930s, for three distinct layers in this top metre of mud in cores from the tropical Atlantic – two layers containing remains corresponding to warm conditions like those in the region today, sandwiching a layer containing remains corresponding to a colder climate. But it was only in 1947 that a Swedish oceanographer, Björe Kullenberg, developed a technique in which a piston in the steel tube is used to suck up the mud while the tube sinks into the sea floor, rather like the way you can suck up water from a bucket using a bicycle pump. The Kullenberg corer made it possible to extract samples of ooze up to 15 metres long, opening wide this window into the climatic history of our planet.

Early studies using the new technique showed that sediment accumulates only very slowly in the wide Pacific Ocean, at a rate of about a millimetre every century. This was good in a way, because it meant that a core 15 metres long covered a large span of geological time (1.5 million years, assuming the deposition rate has stayed constant all that time); but it was bad in another way, because it meant that the cores contained very little detailed information, with even a thousand years of climatic information compressed into a single centimetre of ooze. But Atlantic sediments turned out to be deposited about three times as fast, giving a more detailed insight into past climates, although only for about the past half million years. By 1953, scientists at the Scripps Institution of Oceanography had found evidence for at least nine Ice Ages in the cores. Meanwhile, in the early 1950s a new research group was being established at what became the Lamont-Doherty Geological Observatory, on an estate at Palisades, New York, given to Columbia University by the benefactor Thomas Lamont. Research vessels associated with the Geological Observatory were almost constantly at sea on various surveys, and whatever else they were doing each vessel always obtained a Kullenberg core every day, to be stored, and eventually analysed, back in Palisades.

The immediate beneficiaries of this cornucopia were David Ericson and his assistant Goesta Wollin. Wollin was a particularly determined individual who, although a

Swedish citizen, came to America in 1942 and joined the US Army to fight the Nazis, and whose first experience of parachute jumping was to be dropped into Normandy, the night *before* D-Day, as a member of the Intelligence Service of the 82nd Airborne Division; partly because of his wartime service, he was thirty by the time he completed his Master's degree at Columbia in 1952. Ericson, born in 1904, had a more conventional background, being too old to fight in the Second World War, and had been a senior research scientist at Columbia since 1947. By counting the numbers of shells of different species of plankton found in the Atlantic sediments at different depths below the tops of the cores, Ericson, Wollin and their colleagues at Lamont were able to show, by the middle of the 1950s, that there had been an abrupt change in climate 11,000 years ago (as with land-based samples, material this young could be dated using the radiocarbon technique, which is how the accumulation rates were known). Over the next few years, the core samples revealed a pattern of changing climate extending back into the past, but, of course, without any exact dates beyond the range of the radiocarbon technique. In a paper published in the journal *Science* in 1963, Ericson, Wollin and their colleague Maurice Ewing reported the analysis of 'more than 3000 cores raised from all the oceans and adjacent seas during 43 oceanographic expeditions since 1947' from which they had identified 'eight containing a boundary clearly defined by changes in the remains of planktonic organisms'. That boundary, back around 1.5

million years in the past if the sedimentation rate was any guide, represented the onset of the first Ice Age of what is known as the Pleistocene Epoch, the interval of geological time during which modern human beings emerged (some modern estimates set the beginning of the Pleistocene a little earlier, around 1.8 million years ago, but this really represents fine-tuning, and does not affect the broad significance of the discussion). The entire Pleistocene, the sedimentary record showed, was marked by a repeating succession of Ice Ages and Interglacials. In the mid-1960s, Ericson and Wollin presented all the evidence to a wider audience, in a popular book, *The Deep and the Past*; but the evidence was not widely accepted by their colleagues outside Columbia University, partly because other people had reservations about the way the Columbia team interpreted the temperature record from the fossil remains (which depended to some extent on a subjective assessment of the changing planktonic populations), and partly because an independent (and seemingly less subjective) technique using the same sediments seemed to be giving a quite different picture of past climates.

The technique had been developed at the University of Chicago in the late 1940s, and depended on the fact that there are two stable isotopes of oxygen in the air, the common form oxygen-16 and the much rarer form oxygen-18. They both react in the same way chemically (so, for example, they are both present in molecules of water and of carbon dioxide), but one (oxygen-18) is heavier than the other, so that molecules of water (or carbon dioxide) which contain atoms of oxygen-18 are slightly heavier than molecules of water (or whatever) which contain oxygen-16. Some of the oxygen from the water is taken up by living creatures and, in the case of plankton, helps to build their shells; but the proportion of oxygen-18 taken up in this way depends on the temperature of the sea at the time. Creatures that live in colder water have a higher proportion of oxygen-18 in their shells. When they die and their shells fall to the bottom of the sea, the ratio of oxygen-18 to oxygen-16 in the shells is preserved as a kind of fossil temperature record. Once the technique for measuring the isotope abundances had been established, there were many potential applications to be investigated, and in 1950 the opportunity to apply the technique to forams from the sea bed was offered to an Italian graduate student, Cesare Emiliani, who was just finishing his Ph.D.; he had obtained his first degree from the University of Bologna in 1945, and carried out field work in the Apennines before moving to Chicago in 1948. By 1955, Emiliani's analysis of the isotope thermometer from cores obtained in the Caribbean and equatorial Atlantic seemed to have established that there had been seven complete Ice Age–Interglacial cycles in the past 300,000 years alone, and in a paper published that year he drew attention to the resemblance between the pattern of Ice Ages he had found and the Milankovitch cycles. But the pattern found by Emiliani (the new kid on the block) did not match the pattern found by the old hand Ericson at

Columbia, who argued that the isotope variations identified by Emiliani must be caused by some other effect, not by temperature changes. And the suggestion of a link between past temperatures and the Milankovitch cycles annoyed the many geologists who had by now discarded the Milankovitch Model. The disagreement simmered for ten years, with the fact that the two techniques did not agree with one another leading most outsiders to ignore both sets of conclusions. Even a meeting held in New York in 1965 expressly to clear the air failed to resolve the issue – but it did introduce into the debate someone who would play a large part in settling the controversy and finding out just how the isotope thermometer really did work.

John Imbrie was already forty in 1965, a Professor of Geology at Columbia University who had been using statistical techniques to analyse the variations in the populations found in various fossil assemblies. The point of this work was that it tried to use variations involving several different species at once, under the influence of several different outside factors, to unravel at least some of the relationships between the changes in the flora and fauna and the changes in the environment. To take a simple (hypothetical) example, imagine two species living on land in the same region, both of which flourish under cool conditions, but one of which needs plenty of water while the other does not. If both species were in decline, that may be evidence that conditions had got too hot for them; but if one was flourishing well and the other was not, that could indicate a cool temperature combined with a shortage of water. The big criticism of the work by Ericson and his colleagues was that it relied to an undesirable extent on monitoring the changing ocean temperatures from just one species, a foram known as *Globorotalia menardii*. The temperature variations indicated by fluctuations in the *menardii* population matched the temperature fluctuations inferred from the isotope studies for the top two or three metres of the cores, but then the two indicators diverged from one another. But were the *menardii* always responding to temperature changes alone, or were they also sensitive to some other environmental factor – perhaps changes in the salinity of the ocean – which was confusing the picture?

After the 1965 conference, which he attended simply as an onlooker, Imbrie decided to adapt his statistical multi-factor technique and apply it to the whole assemblage of foram species found in the cores, starting out with an intensive study of one particular Caribbean core to prove the technique. It took him three years, working with a student, Nilva Kipp, and during which time he moved from Columbia to Brown University, to develop the technique to provide a record of the changing climate based on an analysis of not one but twenty-five different species of forams found in that core, related not just to the average temperature but to several environmental influences including both summer and winter temperatures and salinity. At the same time, the same core was being analysed in comparable detail using the isotope technique, by Wallace

Broecker and Jan van Donk at Columbia. The results were conclusive. The two analyses agreed with one another, showing that in the earlier debate Emiliani had been essentially right and Ericson had been wrong to rely so much on *menardii* and to assume that all the variations in the *menardii* population were caused by temperature changes. The multi-factor technique showed that changes in salinity of the surface waters of the Caribbean had, indeed, played an important part in influencing the foram population; but it also showed that the temperature fluctuations in the region associated with Ice Ages corresponded to a drop of only 2 °C, whereas Emiliani's analysis had indicated a drop of 6 °C. Something else was amplifying the isotope changes when the Earth cooled.

The obvious candidate for that something else was the way water gets locked up in great ice sheets during an Ice Age. When water evaporates, it is easier for the lighter molecules to escape into the air, so the water left behind tends to have a higher proportion of oxygen-18; much of the evaporated water, relatively rich in oxygen-16 compared with the water left behind (exactly how rich also depends on the temperature), falls as snow during an Ice Age, and gets locked up as ice instead of being recycled back into the sea. So the proportion of oxygen-18 available in the oceans is higher during an Ice Age, even before you take account of the way the proportion of oxygen-18 in their shells is enhanced by the way plankton take up the water. This explanation of why Emiliani's inferred temperature change was too big was just a speculation when Imbrie gave a talk on the subject at a meeting in Paris in September 1969. As he recounts in his book *Ice Ages* (co-written with his daughter, Katherine Palmer Imbrie), his talk was scheduled for four o'clock on the Friday of the meeting:

In Paris, on a warm September afternoon, there are distractions attractive enough to lure even the most dedicated of scientists away from the lecture hall. When [I] finally spoke, it was to an audience of two. Half of the audience understood no English. The other half was Nicholas Shackleton – a young British geophysicist who ... had already published data suggesting that much of the observed isotopic variation reflected changes in the volume of global ice.

Imbrie and Shackleton, each previously in ignorance of the other's work, were delighted to find that they were thinking along the same lines, and both appreciated that far from this discovery being a drawback, if anything it enhanced the power of the isotope technique. More than anything else, it seemed, an enhancement in the oxygen-18 proportion found in the shells of these long-dead plankton was an indication that much of the Earth's water was locked up in ice sheets at the time they died – and the exact isotope ratios even told you how much water was locked up in this way. What better indicator could there be of the advance and retreat of the ice? And, even better, this is a truly global indicator. The plankton population techniques only told you what the temperature was in the Caribbean, or wherever the core came from, at a certain time in the past. The isotope technique, it was now clear, gave you, in effect, a measure of the global average temperature, no matter where in the oceans the core had been drilled. But they still needed a way to date accurately the temperature fluctuations that were now clearly apparent in the cores covering the entire Pleistocene Epoch.

Fortunately, the radioactive dating technique is not unique to carbon-14. In principle, it can be applied to any radioactive isotope of any element, provided that the way in which the radioactive material decays is understood, and the small amounts of residual radioactive material and the products of radioactive decay (the 'daughter' isotopes) can be measured. In the 1950s, and afterwards, physicists developed several different radioactive calendars of this kind, based on the decay of radioactive isotopes of elements such as uranium, thorium and potassium. The half life of one particular radioactive isotope of potassium, known as potassium-40, is 1.28 billion years, while thorium-230 has a half life of 80,000 years, which gives you some idea of the range of dates which can be determined by the various techniques. The thorium calendar, in particular, turned out to be just right for measuring ages on timescales appropriate for testing the validity of the Milankovitch Model. Indeed, it was the application of thorium dating techniques to measure the ages of ancient shorelines on Barbados and New Guinea that helped to revive interest in the Milankovitch Model in the late 1960s.

The fact that sea level had been higher at certain times in the past than it is today had long been known – on his voyage round the world on board HMS *Beagle*, Charles Darwin, who was a geologist before he made a name as a naturalist, was one of several such pioneers who noticed raised beaches, several metres above the present day sea level, in different parts of the world. But until the thorium dating technique was established, working out just when the surf had pounded on those beaches had been largely a matter of guesswork. There is also evidence for former beaches below present day sea level in some parts of the world, indicating that, relative to those levels, the sea level today is high. But in Darwin's day, and right into the twentieth century, it was far from clear whether raised beaches seen today were above sea level because the land they sit on had been uplifted, or because the sea level itself had fallen. In fact, in some cases the land has been uplifted. But if it could be shown that raised beaches the same height above present sea level had formed around the world at the same time, that would be a clear indication that it was the sea itself that had been higher then. And the obvious cause for such changes in sea level is the advance and retreat of ice sheets on land – when more water is locked up in ice, the sea level falls; but when the ice sheets melt, sea level rises.

By the middle of the 1960s, Wallace Broecker and his team at Columbia had used the thorium technique to show that raised beaches in islands of both the Pacific and the

Atlantic oceans corresponded to a period 120,000 years ago when the sea level was six metres higher than it is today; slightly less compelling evidence indicated another, but less extreme, increase in sea level about 80,000 years ago. (All these dates have an element of uncertainty, and should also be regarded as central figures for a span of a few thousand years; a 'date' of 80,000 years before the present should be read as indicating that sea level was high for several thousand years centred around that time, and an age of 80,000 years is regarded as the same as, say, 82,000 years in this context.) This work encouraged other researchers to investigate the raised features carved by ancient waves at sites around the world, and by 1968 it was clearly established that sea level had been high around the world 125,000 years ago, 105,000 years ago and 82,000 years ago, as well as today. These dates correspond to warm periods predicted by the Milankovitch Model, provided that all three of the astronomical effects, and the way they interact with one another, are allowed for. And just at the time all this was going on, in the heartland of Europe another geologist was investigating the layers of wind-blown silt, separated by bands of richer soil, found at different levels below the surface of the ground, and revealed as stripes on the side of a quarry near Brno (the city where the pioneering geneticist Gregor Mendel once lived and worked), in what was then Czechoslovakia.

These layers of fine silt, known as loess, were laid down during Ice Ages, when the region was cold and dry, a desert across which chilly winds blew the fine material. But when the ice retreated and the region warmed up, the rains fell and it became fertile. In 1968 George Kukla was able to show, using yet another new dating technique, that ten cycles of this kind had occurred, with a rhythm 100,000 years long. The icing on the cake came from the latest analysis of the oxygen isotopes from forams in the Caribbean sediment, completed in 1969, in which Broecker and van Donk identified six Ice Age–Interglacial cycles, repeating with a 100,000-year rhythm, and all showing a very rapid switch from Ice Age to Interglacial conditions, with a more leisurely slide back into Ice Age climate. When Broecker and Kukla met in Paris in 1969, at the same scientific gathering where Imbrie and Shackleton got to know one another, they were able to compare notes and discover how well their different pieces of the jigsaw puzzle fitted in to the same big picture. All the evidence fitted together – raised sea levels, the oxygen isotope studies from the ocean sediments, and now the wind-blown soil of central Europe. There really had been a repeating succession of Ice Ages and Interglacials during the Pleistocene, and the dominant rhythm of the Ice Age–Interglacial cycle was the 100,000-year pulsebeat predicted by the Milankovitch Model. The snag was, according to Milankovitch's own calculations the 100,000-year rhythm shouldn't be the dominant factor among the astronomical influences. Why did it show up so much more clearly than the precession and tilt effects? The most likely explanation, as Croll had suggested a century earlier, was that when the Earth's orbit is more elongated, the precession cycle has a greater influence on climate. In which case, more detailed analysis of the climate record ought to show up the precession cycle (and, of course, the tilt cycle) as well as the eccentricity cycle. It was time for a definitive test of that model, and the test would depend crucially on the new dating technique which Kukla had used in his work in Czechoslovakia.

Kukla had not invented this technique, although he was one of the first people to apply it to the study of past climates. It depended upon the discovery that the Earth's magnetic field is not constant, but sometimes (seemingly at random) reverses itself entirely, first fading away to nothing and then building up again in the opposite sense, so that what is now the North magnetic pole becomes the South magnetic pole, and vice versa. The details of exactly how and why this happens are still not known, but it is clearly a result of the way the Earth's magnetic field is generated, by swirling currents of fluid, electrically-conducting, iron-rich material in the deep interior of our planet. All that matters as far as we are concerned, however, is that these magnetic reversals do happen, and that they are preserved as a record in the rocks when flows of molten lava are setting. Each lava flow becomes magnetized as it sets, and retains a permanent imprint of the way the Earth's magnetic field was oriented at the time the lava was being laid down. Studies of the fossil magnetism of the rocks hinted at the possibility of magnetic reversals early in the twentieth century, but it was only in the 1950s that detailed analyses of samples from around the world established this pattern of behaviour. When reversals happen, they take place in less than 10,000 years (perhaps much less), so they show up sharply in the geological record; but once a particular orientation of the field is established, it may last for millions of years, or only for a few tens of thousands of years. The most recent reversal happened about 780,000 years ago, but the Earth's magnetic field is weakening at the moment, so we may be living through the early stages of the next reversal.

The discovery of magnetic reversals was soon put to practical use. First, where the ages of magnetized rocks could be inferred in other ways, geologists were able to put together a magnetic calendar which showed not only in which direction North and South magnetic poles were oriented at different times in the geological past, but also how long each magnetic orientation had lasted. When plotted out on a diagram, the result is rather like the bar codes found on most products purchased in shops today – alternating bands of black (corresponding to present day polarity) and white (corresponding to reversed polarity), showing different patterns of magnetic orientation and their duration. It turned out that often very short intervals of one polarity were embedded within much longer intervals of the opposite polarity, providing very useful markers in time – these short-lived reversals are called 'events' and given the name of the geographical location where they were first discovered; the longer intervals are called 'epochs' and named after appropriate people, such as the people who first identified geomagnetic reversals. With this calendar (or chronology) established, it became possible to date rocks simply in terms of their magnetism – a layer of strata in which a particular pattern of reversals was found could be slotted into place in the chronology by comparing it with the known pattern, like identifying a person from their fingerprints.

The technique has many applications, but what matters here is that the magnetic signal can be read from the cores drilled from the sea bed. With a library of thousands of sea bed cores to draw on, in the second half of the 1960s the Lamont researchers were able to put actual dates on many of the events recorded in those cores, tying them in to an absolute chronology, and then filling in the details between magnetic reversals from other techniques, such as sedimen-tation rates and radioactive chronologies, and the way populations of species such as *menardii* changed from layer to layer in the sediments. It was this kind of magnetic work that, for example, pinned down the date of the start of the Pleistocene to 1.8 million years ago. That particular breakthrough was made by William Berggren, of the Woods Hole Oceanographic Institution, and Jim Hays, from Lamont, who had become the Director of the deep-sea sediments core laboratory in 1967, three years after complet-ing his Ph.D. at Columbia. It was Hays who, in 1970, took the initiative in setting up the team that finally established the validity of the Milankovitch Model.

It was all very well having different teams of physicists, from different backgrounds and disciplines, tackling the Ice Age problem in their own ways, bumping into each other at conferences, and arguing about which technique gave the best insight into the climates of the past. What was really needed was a coordinated effort to combine the various techniques – the chemistry of how plankton absorb oxygen, the biology of which species thrive best under which climatic conditions, the physics of isotope studies, the geology of how sediments were laid down, and so on – into one coherent package. Hays realized that although it would be a daunting task to set up such a research effort from scratch, there were already people looking at the problem in all these different ways at universities and research institutions around the world, and that for a relatively modest investment of time and money their efforts could be coordinated into one super-project. In 1971, he persuaded the US National Science Foundation to fund such a project, under the name CLIMAP, from Climatic Mapping. Its initial objective was to investigate the climate of the North Pacific and North Atlantic oceans over the past 700,000 years, the interval (known as the Brunhes Epoch) since the latest geomagnetic reversal. But this proved so successful that in 1973 the project was extended into an effort to map the distribution of climate zones across the world during the latest Ice Age, and also to determine just how climate had varied during the entire Pleistocene.

The key to the second objective was to find a single sea bed core which contained the best kinds of forams for analysis and which extended, unbroken, all the way back past the most recent magnetic reversal, known as the Brunhes–Matuyama boundary. By chance, it turned out that just such a core (which the researchers referred to as their 'Rosetta Stone', but which officially has the more prosaic label V28-238) had been obtained from the western equatorial Pacific by a Lamont ship in 1971, just when CLIMAP was getting started. Once the importance of this core was realized, Hays sent samples of it for isotopic analysis to Nick Shackleton, at the University of Cambridge, who was by 1971 the acknowledged world leader in the application of this technique to obtain an accurate record of isotope fluctuations from a small amount of material. Shackleton did even better than Hays could have hoped. He not only worked out the isotopic variations from the remains of creatures that lived in the surface waters, obtaining a near-perfect agreement with Emilani's Caribbean data, but also showed that exactly the same pattern of fluctuations occurred in the shells of creatures that spent all their lives in the cold water at the bottom of the Pacific. Since the temperature at the bottom of the sea hardly changes, even during the switch from an Ice Age to an Interglacial, this was the definitive proof that the main influence on the oxygen isotope composition was indeed the advance and retreat of the ice sheets on land, and that the isotopes were recording the pulsebeat of global climate change.

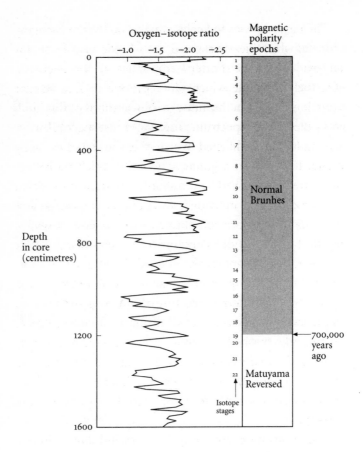

The 'Rosetta Stone' of recent climate variations, showing
magnetic and isotope fluctuations measured by Shackleton and
Opdike from a single sea bed core in 1972 (see text for details;
the illustration is taken from their 1973 paper). This work
established the key link between the isotope variations and the
magnetic reversal 700,000 years ago.

The next step was to analyse this record of the changing climate, which clearly showed the 100,000-year cycle, for any evidence of the shorter astronomical cycles. There is a standard technique which physicists use to find regular variations buried in the 'noise' of observations of this kind; it is called power spectrum analysis (or just spectral analysis), and it can be likened to analysing the sound made by a chord played on a guitar to identify which individual notes are being played by the chord. Any trained musician could pick out the individual notes in this simple case, but the situation is more complicated when what is 'heard' is in effect a discord of many unrelated notes being played together. The geophysicist Edward Bullard once commented to us that the problem is usually more like 'trying to work out the internal structure of a piano by listening to the noise it makes when pushed down a flight of stairs'. Fortunately, the analysis is one of those problems that gets easier and easier as computers get better and better, and Imbrie had already been using the technique in his earlier work, so he was ready to apply it to the data from core V28-238.

The results were tantalizing. The core did show evidence of the presence of cycles roughly 40,000 years long and 20,000 years long, but these 'signals' were not statistically significant – according to standard tests, they could have occurred by chance as a result of random fluctuations. And yet, they *were* the cycles predicted by the Milankovitch Model! Hays and his colleagues realized that V28-238 was not really ideal for this kind of analysis. The sediments in the part of the world where the core was obtained accumulate only at a rate of about a millimetre every hundred years, and the boundaries between the different layers could easily get blurred by the activity of burrowing worms living on the sea floor, smearing out the climatic signal the team was looking for. What they really needed was a comparably long, undisturbed core from a part of the world where the sedimentation rates were much higher. Checking through the Lamont core library, Hays found one, from the Indian Ocean, that was nearly ideal for this purpose. The sedimentation rate in the region was three millimetres per century, and the kinds of planktonic remains preserved in the sediments were just right for analysis; the snag was that it was too short, and extended back only 300,000 years into the past. There was nothing else suitable in the Lamont library, but under the auspices of CLIMAP Hays was able to enquire whether any other laboratories had anything suitable. It turned out that Florida State University had some cores obtained in the same part of the Indian Ocean, including one that went back 450,000 years. The snag with that core was that the top of it had been damaged and broken off when it was extracted from the sea bed. But a large part of the core overlapped with the shorter Lamont core, and where they overlapped the isotopes and other climatic indicators marched precisely in step, like the match between the tree-ring patterns in two samples of wood taken from trees growing in the same forest. Together, the two cores provided a complete record going back nearly half a million years, with an accumulation rate fast enough to ensure that they contained information about cycles as short as 10,000 years.

The 15-metre span of the core was sampled at intervals of 10 centimetres, giving 150 isotope samples spaced 3,000 years apart, which is ample to pick out cycles 20,000 years long using the power spectrum technique. When the spectral analysis was carried out, it clearly showed a statistically significant cycle with a period of 41,000 years, and not one but two shorter cycles, at 23,000 years and 19,000 years. This was a total surprise – but when the Belgian astronomer André Berger was enlisted to check the astronomical calculations (which could now be done quickly and even more accurately than Milankovitch could ever have hoped, using computers) he found that subtleties in the changing gravitational influences of the other planets which had been too detailed for Milankovitch to work out using his faithful fountain pen really did split the '20,000-year' rhythm into two components, with periods of 23,000 years and 19,000 years. Armed with all this information, and steadily improving analysis techniques and computers, the same cycles were soon identified in other climate records. Hays, Imbrie and Shackleton published their findings in the journal *Science* in 1976, under the title 'Variations in the Earth's Orbit: Pacemaker of the Ice Ages':

It is concluded that changes in the Earth's orbital geometry are the fundamental cause of Quaternary ice ages. A model of future

climate based on the observed orbital-climate relationships . . .
predicts that the long-term trend over the next several thousand
years is towards extensive northern-hemisphere glaciation.

The same year that those words were published, the
Milankovitch Model received a particularly significant
endorsement from John Mason, then the Director-General
of the UK Meteorological Office. The significance of this
endorsement was not just due to his position in the
meteorological establishment, but to the fact that this represented a dramatic conversion from someone who had
previously been highly sceptical about the model. Indeed,
the calculations which persuaded Mason that Milankovitch
had been on to something after all were originally carried
out in an attempt to prove that the model would not hold
water, and we were present at the talk Mason gave in 1976
where he astonished his audience with his about face almost
as much as Agassiz astonished his audience, all those years
before in Switzerland, when he talked about Ice Ages rather
than fossil fishes. The final ingredient in making Mason's
talk so dramatic was the simplicity of the physics underlying
his straightforward calculations of the relationship between
the Milankovitch rhythms and the advance and retreat of
ice sheets across the Northern Hemisphere.

The calculations could not have been carried out before,
because it was only in the 1960s and 1970s that a good
knowledge of just how the ice sheets had advanced and
retreated over the past 100,000 years had been established.

But the calculations need no high-speed electronic computers, and could literally have been carried out on the back of the proverbial envelope, had Mason been so inclined. They depend simply on the amount of heat which is required to turn ice at 0 °C into water at the same temperature – the latent heat of fusion, which is (in the units used by Mason) 80 calories for every gram of ice melted. Since one calorie is defined as the amount of heat needed to raise the temperature of one gram of water by one degree Celsius, this means that the heat required to melt one gram of water at the freezing point is enough to heat that same gram of liquid water all the way from 0 °C to 80 °C. When you are melting glaciers, that adds up to a lot of heat, which is why Mason started the calculation expecting to prove that the change in heat balance of the Northern Hemisphere caused by the astronomical rhythms would not be sufficient for the task.

A similar process operates in reverse when water vapour condenses into liquid or water freezes into ice. In each case, latent heat is given out by the water, rather than being taken up. When the vapour condenses into water at the same temperature, 595 calories of heat are released for each gram involved; so when the vapour goes all the way to the solid form and falls as snow, it liberates 675 calories for every gram of snow that falls. This heat goes into warming the surrounding air and the globe generally, while the need for heat to be absorbed in melting snow and ice tends to keep regions covered by winter snow cool well into early summer. Each year, we see the Ice Age cycle repeated in miniature. Mason calculated that between 83,000 years ago and 18,000 years ago, essentially the extent of the latest Ice Age, the overall deficiency in insolation in the crucial region of the Northern Hemisphere, according to Milankovitch's own figures, was 4.5×10^{25} calories. Comparing this with the amount of ice that spread across the region, he found a deficit of 1,000 calories for every gram of ice that had formed, very close to (but, significantly, a little greater than) the 675 calories that had to be got rid of to turn that much water vapour into ice. From 18,000 years ago until the present day, when the astronomical cycles conspired to increase the summer warmth of the crucial Northern Hemisphere region, the excess of insolation was 4.2×10^{24} calories, which compares with 3.2×10^{24} calories required to melt the volume of ice known to have melted during this interval. This is a truly dramatic agreement between astronomically large numbers, which stretches across 24 'orders of magnitude', or powers of ten (to put it in persective, 10^{24} is roughly equal to the number of bright stars in all the known galaxies in the entire Universe). The two numbers differ by only 1 part in 10^{24}, equivalent to a decimal point followed by 23 zeroes and a 1. No wonder Mason (and his audience) found this persuasive evidence that Milankovitch was right after all.

The escape from Ice Age conditions, beginning 18,000 years ago, required the combined influence of all three astronomical cycles to drag the Earth into a peak of warmth about 6,000 years ago. Orbital eccentricity changes combined with a shift in the wobble of the Earth which made June the month of closest approach to the Sun, boosting the heat of Northern Hemisphere summers, just at a time when the tilt of our planet reached a maximum, putting that summer Sun particularly high in the sky. Since 6,000 years ago, all these factors have turned around, and conditions for Northern Hemisphere summer warmth are becoming less favourable. The prospect is for a return of Northern Hemisphere glaciation, on a timescale of thousands of years.

But not just Northern Hemisphere glaciation. One of the initially surprising fruits of the isotope studies was the discovery that the global climate changes in step in both hemispheres, with Ice Ages occurring simultaneously in both hemispheres – if not, the isotope changes produced by growing ice sheets in one hemisphere would be cancelled out by the effects of shrinking ice sheets in the other hemisphere, and the astronomical rhythms would not show up in this way. It seems likely that the reason for this is that the overall climate is driven by what happens in the Northern Hemisphere (which is particularly sensitive to the astronomical influences, for reasons we have already explained), with the addition of feedbacks, such as changes in the ocean currents of the kind discussed originally by Croll. Another way of looking at this is to start from the fact that the conditions which are required to melt snow and ice on land in the Northern Hemisphere (warm summers, causing an Interglacial) go hand in hand with relatively warm *winters* in the Southern Hemisphere, and this discourages the formation of extensive sea ice. The land of Antarctica is always covered by ice, even during an Interglacial, and particularly *cold* winters are what you need to freeze the top layer of the oceans. But there is no need to understand all of the subtleties of how the various cycles work in order to appreciate that they do work (the sea bed sediments prove that), and to use them to gain insight into our own origins.

Epilogue

Ice Ages and Us

People are mammals – the most successful mammals, in terms of their distribution around the world and the way they manipulate their environment, that have ever lived. Mammals, in turn, are among the most successful forms of life on Earth today. But the dominance of mammals, and of ourselves, in the world today is a direct result of the death of the dinosaurs, some 65 million years ago. The fossil evidence shows that before then our ancestors were small, scurrying creatures that lived literally in the shadow of the dinosaurs, and all the large-animal niches in the ecology, occupied today by creatures such as giraffes, elephants, apes, rhinoceros, wildebeest, lions and so on, were occupied by dinosaurs. Mammals were only able to spread out to occupy those niches, diversifying into the variety we see today, after the dinosaurs had gone. The best evidence we have is that the dinosaurs were already in decline by 65 million years ago (possibly because the Earth experienced a very dry phase, caused by the geographical distribution of the continents at that time) when they were finished off by the environmental upheaval associated with the impact of an object from space about 10 kilometres across at what is now the Yucatan Peninsula of Mexico. Only small creatures that could easily find shelter and needed little food were able to survive in the millennia that followed the disaster.

The geological evidence, now well-dated using the techniques we have described, shows that during the 60 million years or so from the death of the dinosaurs to the beginning of the present Ice Epoch (the time during which mammals were diversifying to fill all those niches vacated by the dinosaurs), the temperature of our planet experienced a slow, somewhat erratic decline, as a result of the way the continents were shifting about on the surface of the globe, altering the flow of ocean currents and the way sunlight was absorbed and reflected back into space. This sounds like a long span of time, and it is, by human standards. But 60 million years represents only 0.13 per cent of the history of the Earth so far, and only a third of the span of time during which the dinosaurs dominated the planet. Geologists give names to the various spans of past time in accordance with the changes they see in the geological record; this provides them with a relative chronology, establishing which events came earlier and which later in the history of the Earth, which can then be tied in to absolute dates as suitable dating techniques become available. We shall pick up the story towards the end of the geological epoch known as the Miocene, now dated as lasting from about 24 million years ago (24 Myr BP, for 'Before Present') to about 5 Myr BP. This was followed by the Pliocene, which lasted only until 1.8 Myr BP, and then by the Pleistocene. There is a tendency for the lengths of named epochs to get shorter as we come closer to the present day, chiefly because geologists can see more detail in the recent record, picking out more subtle changes, not necessarily because the pace of geological change is actually hotting up. In the most extreme example of this chauvinistic approach, geologists set the start of the present epoch, the Holocene, at the beginning of the present Interglacial, 10,000 years ago. This is completely unjustified, since there is no evidence that the present Interglacial marks the end of the Ice Epoch that has persisted for the past few million years; the boundary is really chosen to mark the emergence of human civilization, as much out of hubris as chauvinism. But we will not be concerned here with anything that happened as recently as 10,000 years ago.

Although there had been other Ice Epochs in the distant geological past, as far as we can tell the Earth had been essentially ice-free for tens of millions of years during the rise of the mammals following the death of the dinosaurs. That began to change in the middle of the Miocene, as Antarctica slowly drifted across the South Pole, and took up more or less the position it occupies today. Nick Shackleton, from his studies of sea-floor sediments, found that ice sheets began to grow in East Antarctica about 13 million years ago, and there is other evidence that by about 10 million years ago there were small glaciers on the mountains of Alaska. Around 6 Myr BP, as Australia and South

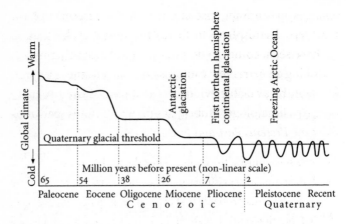

The downward slide of global temperature since the Paleocene, 65 million years ago. Note the horizontal time scale is not linear, so that more recent variations, largely the result of the Milankovitch process, show up more clearly. (Modified from J. Andrews, in *Winters of the World*, B.S. John (ed.), Newton Abbot: David & Charles, 1979).

America moved away from Antarctica, the circumpolar current which flows right round Antarctica had nothing to obstruct its flow, and intensified, locking the south polar region into a pattern of intense cold that has persisted, with some fluctuations in the volume of ice, right up to the present day. With the Southern Hemisphere locked into this pattern, as we come closer up to date and the continents in the Northern Hemisphere shift into their present positions, cutting off the flow of warm water to the Arctic Ocean and making the region particularly sensitive to changes in summer insolation, it is the response of the Northern Hemisphere to the astronomical cycles that, as we have seen, comes to dominate the Ice Age rhythms.

Taking an extremely cautious approach to the interpretation of the available evidence, the doyen of late twentieth century climatology, Hubert Lamb, wrote in his epic book *Climate: Present, Past and Future*:

At some stage during the cooling after the mid Miocene great ice sheets did come into existence in Antarctica, certainly by 7 million years ago. It seems probable that the first continent-wide ice sheet formed on the central plateau of East Antarctica, which was probably 2500 to 3000 m above sea level before the ice load reduced its level by isostatic adjustment to the present height of the subglacial bedrock (which still exceeds 2000 m above sea level in the heart of East Antarctica). It was probably only after this great ice sheet was there that the climate could become severe enough for the glaciers in the high mountains of West Antarctica to produce thickening and coalescing ice-shelves on the deep ocean among the islands of what was then an archipelago, and for these ice shelves to become grounded and ultimately fill the ocean deep in that area. The West Antarctic ice sheet also existed by 7 million years ago.

The 'isostatic adjustment' Lamb refers to is the way the solid rock of the Earth's crust sinks into the fluid material below under the weight of the ice, and rebounds when that weight is released; this also happened in Scotland during the most recent Ice Age, pivoting Britain like a see-saw so that the southern part of the island rose up. Since the ice melted, Scotland has been rebounding from this depression, and southeast England is still, as a result, sinking slowly relative to sea level even after 10,000 years.

It was at about the time of these major glaciations in Antarctica that the geography of the Northern Hemisphere began to resemble closely that of the present day. South America, moving northward, gradually caught up with North America, so that by about 3 Myr BP the gap between them was closed, and ocean currents that used to flow westward through that gap were being diverted northward as the Gulf Stream, setting up the pattern of circulating ocean currents that we see today. But the drifting continents were also closing the gaps around the Arctic Ocean, so that this northward flow of warm water could not penetrate all the way into the polar sea. The first Northern Hemisphere glaciation of the present Ice Epoch, dated using the radioactive potassium method, occurred about 3.6 million years ago. This was a particularly significant event in the evolution of humankind, because fossil remains show that our ancestors lived in East Africa at that time. It wasn't so much the cooling itself that affected them, as the fact that, during an Ice Age, with lowered temperatures there is less evaporation from the oceans, and therefore less rainfall. Together with changes in the pattern of circulation of the atmosphere caused by the presence of ice sheets at higher latitudes, this means that with the present geography of the globe when Europe experiences an Ice Age, East Africa experiences a drought. So the forest in which our ancestors lived shrank when the ice advanced.

There is independent evidence that something dramatic happened to our ancestors at the time the current Ice Epoch began. By comparing the DNA of closely related species, molecular biologists can tell how long it is since the species shared a common ancestor. Studies of species in which the split from a common ancestor can be dated accurately from fossils show that changes in the DNA accumulate at a steady rate. The DNA of human beings differs from the DNA of the chimpanzee and the gorilla by about the same amount as they differ from each other, a little over one per cent (so, in genetic terms, we are about as closely related to a gorilla as a chimpanzee is). The difference between our DNA and the DNA of the orang-utan, however, is about twice as great. This tells us immediately that the evolutionary line which has produced both ourselves and the chimp and gorilla split off from the evolutionary line leading to the orang-utan twice as long ago as the almost simultaneous three-way split which gave rise to people, chimps and gorillas.

As with geological epochs, it is straightforward to determine the relative chronology, and, again as with geological time intervals, these relative chronologies can be tied in to absolute dates by other techniques – in this case, using fossils, which are themselves dated from the geological record. Put all of the evidence together, and it tells us that a forest-dwelling East African proto-ape line gave rise to three separate lines, leading to ourselves, the chimpanzees and the gorillas, between about 3.5 and 4 million years ago, exactly when the climate was changing dramatically. Since both the Ice Ages and the evolutionary changes are tied to the same absolute timescale (ultimately, through radioactive potassium), there is no doubt that the evolutionary changes and the environmental changes occurred at the same time. Conceivably (but highly improbably!) the geological timescale might be adjusted once again; but if it is, the evolutionary timescale will change in step with it. It is hard to escape the conclusion that the changes in the environment in which our ancestors lived were responsible for the three-way split, and it is straightforward to think of a model of how the repeating succession of Ice Ages and Interglacials that followed, repeating with the astronomical rhythms calculated by Milankovitch and found in the sea-bed record by Hays, Imbrie and Shackleton, could have encouraged the emergence of a species like ourselves from the raw material of a proto-ape.

The distinguishing characteristic of human beings is versatility. Some animals run faster, some are better swimmers, some have better teeth and claws for killing and eating meat, some have better teeth and digestive systems for eating plants, and so on. But people do a little bit of everything quite well. Think of what happens to a forest dwelling, apelike creature when the forests shrink. Resources become scarce, and there is fierce competition for them, creating an evolutionary pressure on the apes (let's call them that) in the forest to become better adapted to a life in the trees. 'Competition', in the Darwinian sense, doesn't mean that the apes fight each other for food, or a safe place to sleep, or whatever. It means that some individual apes are better than others at, say, climbing the branches to get fruit. Those individuals will survive better, because they are well fed, and will produce more offspring, who will inherit the parent's agility at climbing about the branches. Less well-adapted individuals have fewer descendants, and over the generations the species evolves, with every individual in later generations being better adapted to their way of life than their distant ancestors were.

But what happens, under those circumstances, to the less successful individuals? They don't just curl up and die (or, if they do, they don't leave any descendants). In our example, those that don't die will be pushed to the edge of the forest, where they will scavenge a living as best they can, in spite of being less agile than their cousins. They will have to try different kinds of food, or starve; and they will have to get about as best they can on the savanna fringing the forests, perhaps running back into the trees for cover when well-adapted plains dwellers approach. There is, indeed, a selection pressure on the edge of the forest that encourages versatility, just as there is a selection pressure in the heart of the forest that encourages specialization. Of course, it is impossible to adapt to a new lifestyle in a few generations, and the population of these less successful apes will decline dramatically, winnowing out the really hopeless ones and leaving just a few survivors, the best of a bad lot, who are the most versatile of a pretty pathetic bunch. If the drought had continued indefinitely (or even for a million years), they would surely all have died out. But after 100,000 years or so, the ice in the north retreated as an Interglacial set in, and the rains returned to East Africa. In a time of plenty, even the surviving groups of less successful apes on the fringe of the forest could find food in abundance, and their population would have soared, while the specialist apes in the heart of the forest were largely unaffected by the climate change, except that they had more forest to occupy. By the time the next Ice Age and its accompanying drought occurred, there would have been a large population of proto-humans (let's call them that), slightly more versatile than their ancestors had been, on which the whole process could operate again. The whole cycle then repeated another ten or a dozen times in each million years. It seems likely that it was this repeated winnowing of the population, followed each time by a respite which allowed the indi-viduals selected by evolution to prosper temporarily, that encouraged the emergence of a species with all the human characteristics of versatility that we have described – and, of course, it is easy to see how intelligence would also be an advantage in a changing world. There is no absolute proof that this is how we evolved from the same ancestor that produced the chimp and the gorilla. Without a time machine, there never will be. But it is the best explanation available, and one that has widespread support. Without the astronomical rhythms of the Ice Ages, we would probably still be tree-apes. It was the repeated drying out and recovery of the East African forests that pushed our ancestors out on to the plains, forced them to become more versatile, encouraged them to walk upright rather than climbing on branches, and, almost as an afterthought, made us intelligent. Fully modern humans, *Homo sapiens sapiens*, emerged during the previous Interglacial to our own, by about 100,000 years ago, and had just one more Ice Age to endure before they began to build civilization. We are the product of the latest Ice Epoch, in a way that Agassiz, Croll and Milankovitch could never have guessed, and that realization is the ultimate triumph of the theory of Ice Ages.

Sources

Joseph Adhémar, *Révolutions de la mer*, published privately by the author (Paris: 1842).

Elizabeth Cary Agassiz, *Louis Agassiz, his life and correspondence*, Houghton Mifflin & Co, Boston, 1886 (published in two volumes).

Louis Agassiz, 'Discours prononcé à l'ouverture des séances de la Société Helvétique des Sciences naturelles', *Actes de la Société Helvétique des Sciences naturelles*, vol. II, pp. v–xxxii, 1837.

Louis Agassiz, *Études sur les glaciers*, published privately by the author (Neuchâtel: 1840).

Louis Agassiz, 'On glaciers and the evidence of their having once existed in Scotland, Ireland and England', *Proceedings of the Geological Society of London*, vol. 3, pp. 327–32, 1840.*

André Berger, *Nature*, vol. 269, pp. 44–5, 1977.

---

* This printed volume actually covers the years 1838–42, but the paper was presented in 1840, as described in the text; the same applies to the papers by Buckland and Lyell from the same volume, also cited here.

A. Berger, J. Imbrie, J. Hays, G. Kukla and B. Saltzman, eds., *Milankovitch and Climate*, Part 1, Reidel, Dordrecht, 1984.

W. Broecker and J. van Donk, *Reviews of Geophysics and Space Physics*, vol. 8, p. 169, 1970.

William Buckland, 'Memoir on the evidence of Glaciers in Scotland and North of England', *Proceedings of the Geological Society of London*, vol. 3, pp. 332–7, 1840.

James Croll, 'On the physical cause of the change of climate during geological epochs', *Philosophical Magazine*, vol. 28, pp. 121–37, 1864.

James Croll, *Climate and Time in their Geological Relations*, Daldy, Isbister & Co, London, 1875.

C. Emiliani, *Journal of Geology*, vol. 63, p. 538, 1955.

D. B. Ericson, M. Ewing and G. Wollin, *Science*, vol. 139, p. 727, 1963.

David B. Ericson and Goesta Wollin, *The Deep and the Past*, Knopf, New York, 1964.

James Geike, *The Great Ice Age*, Isbister, London, 1874.

C. M. Goodess, J. P. Palutikof and T. D. Davies, *The Nature and Causes of Climate Change*, Belhaven, London, 1992.

J. D. Hays, J. Imbrie and N. J. Shackleton, *Science*, vol. 194, pp. 1121–32.

Arthur Holmes, *Principles of Physical Geology*, Nelson, London, 1944; revised edition 1965.

John Imbrie and Katherine Palmer Imbrie, *Ice Ages: Solving the Mystery*, Macmillan, London, 1979.

James Irons, *Autobiographical sketch of James Croll, with memoir of his life and work*, Stanford, London, 1896.

W. Köppen and A. Wegener, *Die Klimate der Geologischen Vorzeit*, Borntraeger, Berlin, 1924.

H. H. Lamb, *Climate: Present, Past and Future*, vol. 1, Methuen, London, 1972.

H. H. Lamb, *Climate: Present, Past and Future*, vol. II, Methuen, London, 1977.

E. Lurie, *Louis Agassiz*, University of Chicago Press, 1960.

Charles Lyell, *Principles of Geology*, Penguin, London, 1997 (originally published in three volumes by John Murray, London, 1830–33).

Charles Lyell, 'On the Geological Evidence of the Former Existence of Glaciers in Forfarshire', *Proceedings of the Geological Society of London*, vol. 3, pp. 337–45, 1840.

Charles Lyell, *Elements of Geology*, John Murray, London, 1865.

Milutin Milankovitch, *Théorie mathématique des phénomènes thermiques produits par la radiation solaire*, Gauthier-Villars, Paris, 1920.

Milutin Milankovitch, *Durch ferne Welten und Zeiten*, Koehler & Amalung, Leipzig, 1936.

Milutin Milankovitch, *Kanon der Erdbestrahlung und seine Andwendung auf das Eiszeitenproblem*, Royal Serbian Academy, Belgrade, 1941. The Israel Program for Scientific Translations published an English translation of this work (*Canon of Insolation and the Ice Age Problem*) in 1969.

N. Shackleton, *Nature*, vol. 215, p. 15, 1967.

Tom Standage, *The Neptune File*, Allen Lane, London, 2000.

Leonard Wilson, *Charles Lyell*, Yale University Press, New Haven, 1972.